Contemporary's

NUMBER POWER 3

Algebra

ROBERT MITCHELL

CONTEMPORARY BOOKS
a division of NTC/Contemporary Publishing Group
Lincolnwood, Illinois USA

ISBN: 0-8092-5714-9

Published by Contemporary Books,
a division of NTC/Contemporary Publishing Group, Inc.,
4255 West Touhy Avenue,
Lincolnwood (Chicago), Illinois 60646-1975 U.S.A.

Manufactured in the United States of America.

9 0 GB(H) 37 36 35 34

Production Editor
Gerry Lynch

Illustrations
Ophelia Chambliss-Jones

TABLE OF CONTENTS

TO THE STUDENT

Welcome to algebra:

Algebra is a powerful tool that makes solving complicated problems much easier. It is the basic mathematical language of technical fields from auto mechanics to nursing and electronics. Also, algebra is a standard section on almost all educational and vocational tests, including GED, college entrance, civil service, and military tests.

This workbook is designed to prepare you for taking a test and for pursuing further education or training that requires basic algebra. The first part of the book, BUILDING NUMBER POWER, provides step-by-step instruction in the fundamentals of algebra. This part is divided into six chapters, and each is followed by a Skills Inventory that you can use to check your progress in the chapter. If you are already familiar with material in a chapter, you can turn to the Skills Inventory to identify those sections of the chapter you should review. However, because algebra is such an important part of your mathematics education, we recommend that you sharpen all your skills by working every problem in the book, even in those sections that you feel you already know.

The second part of the book, USING NUMBER POWER, will give you a chance to apply algebraic skills in real-life situations. These applications are fun and will give you a chance to see the power of algebra.

To get the most out of your work, do each problem carefully and check each answer to make sure you are working accurately. An answer key is provided at the back of the book.

BUILDING NUMBER POWER

INTRODUCTION TO ALGEBRA

Students often ask, "what is algebra?" In this book, you will learn that algebra, like arithmetic, is a tool that uses addition, subtraction, multiplication, and division to solve problems. These two pages contain a brief introduction to four concepts that are especially important in the study of algebra.

Algebra uses both positive and negative numbers.

In arithmetic, you learned to add, subtract, multiply, and divide numbers greater than 0, called positive numbers in algebra. In this book, you'll learn these same operations for negative numbers. A negative number is a number that is less than 0. Indicate a negative number with a negative sign (−). A positive number is commonly written without any sign.

EXAMPLE: Positive numbers: 1, 2, 3, 4, 5
Negative numbers: −1, −2, −3, −4, −5

1. Check Your Understanding: Write the numbers that represent 6 above zero, 6 below zero, 20 above zero, and 20 below zero.

________ ________ ________ ________

You'll learn more about positive and negative numbers in chapters called SIGNED NUMBERS and POWERS AND ROOTS.

Algebra uses letters to stand for numbers.

Suppose you want to represent the word expression "some number plus 5" by symbols. Represent "some number" by a letter. Any letter can be used.

EXAMPLE: "Some number plus 5" can be written "$x + 5$."

The expression "$x + 5$" is called an *algebraic expression*. To find a value for an algebraic expression, *substitute* (replace) the letter with a number and then do the indicated arithmetic.

EXAMPLE: What is the value of $x + 5$ when $x = 4$?
Replacing x with 4, the expression becomes $4 + 5$, which is 9.
Thus, for $x = 4$, the value of $x + 5$ is 9.

2. Check Your Understanding: Find the value of each algebraic expression below by substituting the given value for the letter.

$x + 3$, for $x = 8$ $y + 4$, for $y = 5$ $a - 7$, for $a = 9$

You'll have a chance to increase your skills in evaluating more complicated algebraic expressions and formulas in ALGEBRAIC EXPRESSIONS.

In algebra, problems to be solved are represented as equations.

In many algebraic problems, you try to find an unknown number. For example, you might want to know what number is added to 5 to get a sum of 11. In arithmetic, you might write the problem in one of several possible ways:

$$\_\_\_ + 5 = 11 \quad \textit{or} \quad ? + 5 = 11 \quad \textit{or} \quad \square + 5 = 11$$

In algebra, you represent the missing number by a letter. You call this letter the ***unknown*** because it represents a number whose value you do not yet know. As an algebraic equation, the above problem can be written as follows:

$$x + 5 = 11 \quad \textit{or} \quad y + 5 = 11 \quad \textit{or} \quad n + 5 = 11$$

3. Check Your Understanding: Rewrite the following problems as algebraic equations, using letters in place of other symbols.

$\_\_\_ + 7 = 13$ $\qquad ? + 6 = 9$ $\qquad \square - 8 = 23$ $\qquad ? - 5 = 10$

An algebraic equation is solved by finding the value of the unknown that makes the equation a true statement.

In the equation $x + 5 = 11$, there is only one value for x that makes the equation a true statement. For example, try two different values for x: $x = 4$ and $x = 6$.

Try $x = 4$: Does $4 + 5 = 11$? No, $4 + 5 = 9$

Try $x = 6$: Does $6 + 5 = 11$? Yes.

As you see, the value $x = 4$ *does not* solve the equation because $4 + 5$ does not equal 11. However, the value $x = 6$ *does* solve the equation because $6 + 5$ does equal 11. The value $x = 6$ is called the ***solution*** of the equation $x + 5 = 11$.

4. Check Your Understanding: Solve each algebraic equation below by finding the value of the unknown that makes the equation a true statement.

$x + 3 = 5$ $\qquad y - 4 = 6$ $\qquad z$ times $2 = 18$ $\qquad n$ divided by $2 = 9$

Much of your work in algebra will be learning to write and solve algebraic equations. As equations get more complicated, your skills will increase. You'll learn more about equations in the chapters called EQUATIONS and RECTANGULAR COORDINATES.

Now, check your answers and correct any mistakes. Turn to the first section of the book, SIGNED NUMBERS, to begin your study of algebra.

Answers to Check Your Understanding:

1. 6, −6, 20, −20
2. 11, 9, 2
3. Answers will differ depending on the letters used. Possible answers include: $x + 7 = 13$, $y + 6 = 9$, $a - 8 = 23$, $b - 5 = 10$
4. $x = 2$, $y = 10$, $z = 9$, $n = 18$

SIGNED NUMBERS

INTRODUCING SIGNED NUMBERS

The study of algebra begins with the study of numbers.

Look at the thermometer to the right. Positive and negative signs are used to indicate temperatures above and below zero (0):

Numbers with a positive sign (+) are called *positive numbers*. Positive numbers represent temperatures greater than (above) zero.

Numbers with a negative sign (−) are called *negative numbers*. Negative numbers represent temperatures less than (below) zero.

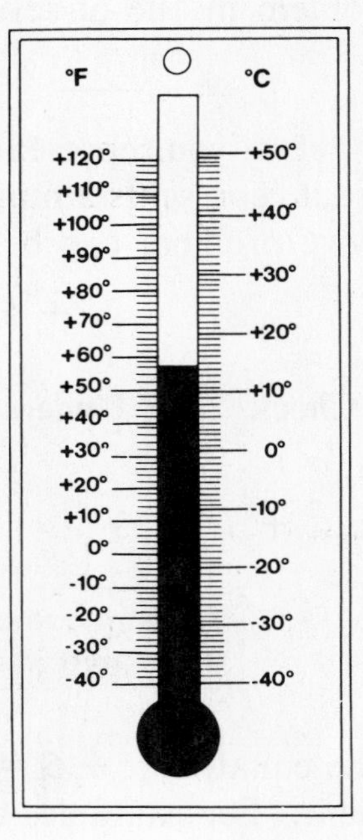

Positive and negative numbers are called *signed numbers*. The thermometer is a familiar example of the use of signed numbers.

Signed numbers are often used to indicate opposite quantities. Other familiar examples include distances above and below sea level and gains or losses in stock market prices.

Notice on the thermometer that the number zero (0) separates positive numbers from negative numbers. Zero itself is neither positive nor negative.

Positive numbers can be written either with a positive sign (+) or with no sign at all. Negative numbers are always written with a negative sign (−).

Positive 20 is written as $+20$ or 20 Negative 20 is written as -20

Positive $5\frac{3}{4}$ is written as $+5\frac{3}{4}$ or $5\frac{3}{4}$ Negative 7.5 is written as -7.5

Zero has no sign and is always written as 0

Remember: *A number written without a sign is always a positive number.*

Write each number as a signed number with a + sign or with a − sign.

1. positive 8 = positive 3 = positive 12 = positive 7 =

2. negative 5 = negative 4 = negative 19 = negative 8 =

3. positive $2\frac{1}{2}$ = positive $4\frac{3}{4}$ = negative $\frac{2}{3}$ = negative $\frac{1}{4}$ =

Identify each number as either "positive" or "negative."

4. $+8$ -7 13 -6

5. $2\frac{2}{3}$ $+4\frac{1}{2}$ -9.3 5.6

THE NUMBER LINE

Signed numbers are often written on a *number line*. Positive numbers are written to the right of zero (0), and negative numbers are written to the left. Notice that positive numbers increase as you move toward the right and that negative numbers increase as you move in the opposite direction—toward the left.

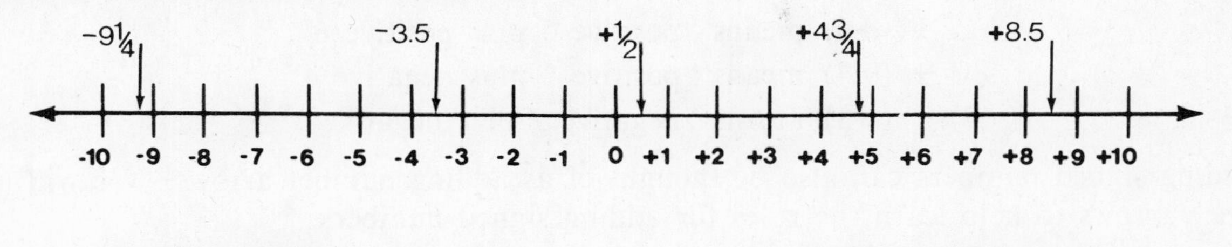

negative numbers (written with a negative sign)

positive numbers (written with a positive sign or no sign)

Number arrows are often used with a number line to represent signed numbers. The length of the arrow represents the size of the number.

A positive number arrow represents a positive number and points to the right. A negative number arrow represents a negative number and points to the left.

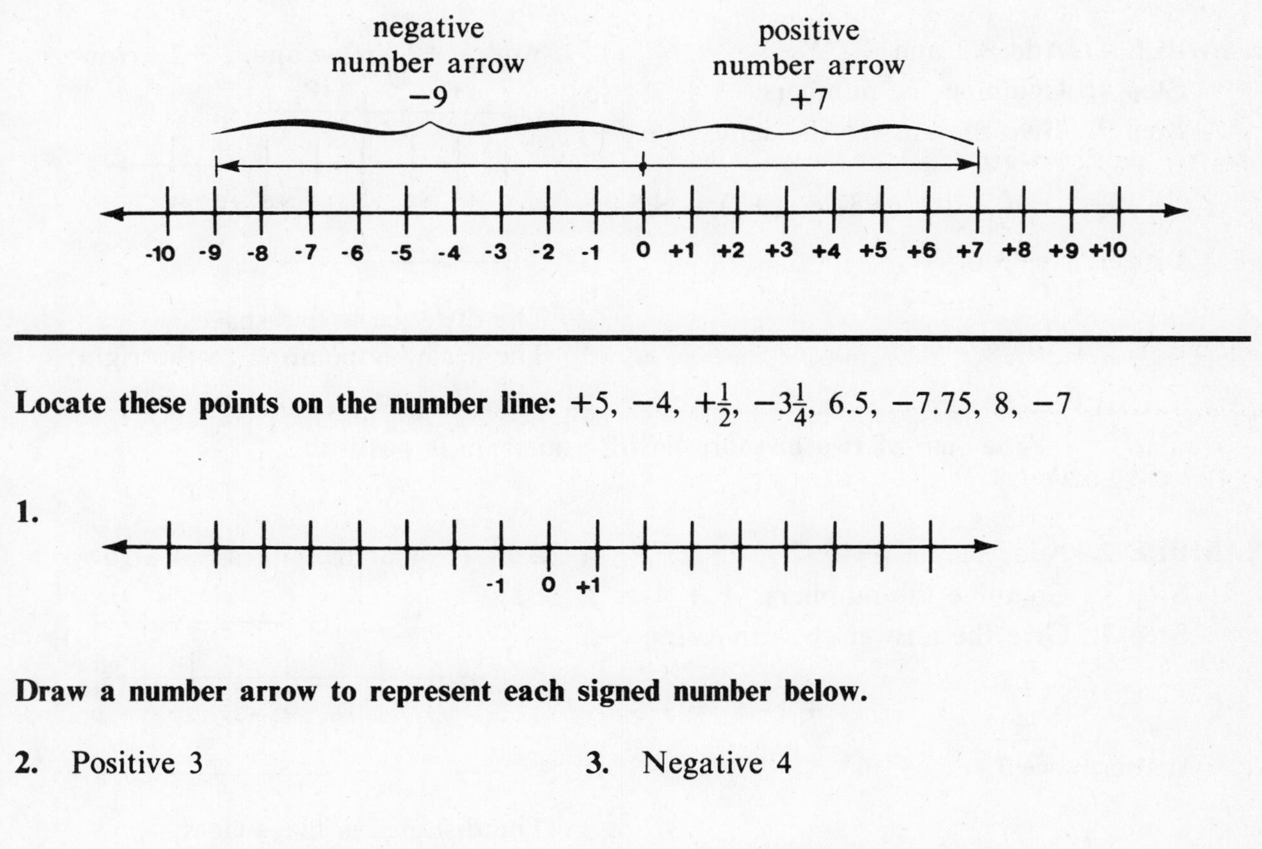

Locate these points on the number line: $+5$, -4, $+\frac{1}{2}$, $-3\frac{1}{4}$, 6.5, -7.75, 8, -7

1.

-1 0 +1

Draw a number arrow to represent each signed number below.

2. Positive 3

3. Negative 4

ADDING SIGNED NUMBERS

In arithmetic, you learned that a positive sign (+) means addition and a negative sign (−) means subtraction. You now know that a positive sign can also indicate a positive number and a negative sign can indicate a negative number. To help avoid confusion, a signed number is often enclosed in parentheses:

+ 5 + (+6) means "positive 5 plus positive 6"
3 + (−4) means "positive 3 plus negative 4"
(−2) + (−3) means "negative 2 plus negative 3"

Adding signed numbers can also be thought of as adding number arrows. We will use number arrows to help learn the rules for adding signed numbers.

RULE 1: To add two or more numbers that have the same sign, combine the numbers and give the answer that sign.

Notice in the examples below that an addition problem can be written either vertically (up and down) or horizontally (across).

EXAMPLE 1. Add +3 and +2.

Step 1. Combine the numbers. 3 + 2 = 5
Step 2. Give the answer the same sign (+).

$$\begin{array}{r} +3 \\ \underline{+2} \end{array} \quad \textit{or} \quad (+3) + (+2) = +5$$

Answer: **+5** or **5**

Add a +3 arrow and a +2 arrow:

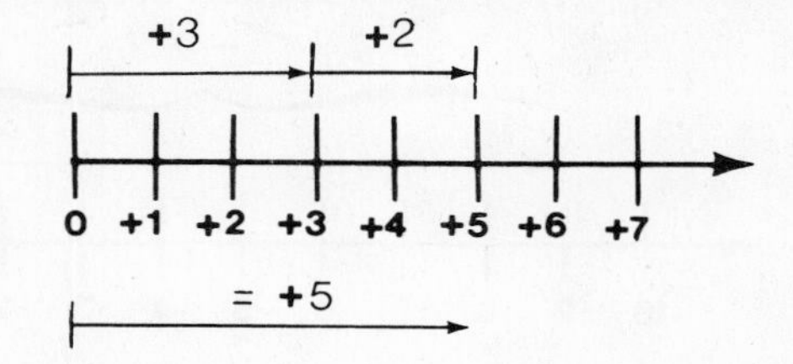

The distance is five spaces.
The arrow is pointing to the right.

The sum of two or more positive numbers is positive.

EXAMPLE 2. Add −1 and −4.

Step 1. Combine the numbers. 1 + 4 = 5
Step 2. Give the answer the same sign (−).

$$\begin{array}{r} -1 \\ \underline{-4} \end{array} \quad \textit{or} \quad (-1) + (-4) = -5$$

Answer: **−5**

Add a −1 arrow and a −4 arrow:

−4 + −1
-7 -6 -5 -4 -3 -2 -1 0
= −5

The distance is five spaces.
The arrow is pointing to the left.

The sum of two or more negative numbers is negative.

Using Rule 1, add the following positive numbers.

Examples

7	**1.** 9	14	$25\frac{1}{2}$	13	$11\frac{1}{4}$
4	6	9	8	12	$8\frac{1}{2}$
11					

+34	**2.** 27	53	+164	$+345\frac{5}{12}$	629.37
+28	53	39	+98	$+453\frac{1}{12}$	537.19
+62					

24	**3.** +16	43	132	+235	315
14	+9	21	213	+129	98
32	+11	18	195	+73	148
70					

3 + (+4) = 7

24 + (+35) = 59

13 + (+21) + (+15) = 49

4. 8 + (+9) = 12 + 9 =

5. 76 + (+83) = 75 + 39 =

6. 35 + (+8) + (+12) = 35 + 18 + 37 =

Using Rule 1, add the following negative numbers.

Examples

−9	**7.** −7	−8	−5	$-17\frac{2}{5}$	−29.3
−8	−9	−3	−6	$-9\frac{1}{5}$	−7.2
−17					

−18	**8.** −24	−25	−83	−172	−218
−9	−10	−31	−26	−53	−100
−8	−9	−43	−39	−182	−296
−35					

−8 + (−5) = −13

(−9) + (−6) + (−7) = −22

9. −9 + (−4) = −12 + (−23) =

10. (−12) + (−9) + (−7) = −12 + (−23) + (−19) =

RULE 2: **To add a positive number and a negative number, find the difference between the two numbers and give the answer the sign of the larger number.**

As the following examples show, the sum of a positive number and a negative number may be positive, negative, or zero.

EXAMPLE 1. Add +5 and −2.

Step 1. Use the numbers without their signs. Find the difference between 5 and 2.
$5 - 2 = 3$

Step 2. Give the answer (3) the same sign (+) of the larger number (5).

Add: $\begin{array}{r} +5 \\ \underline{-2} \end{array}$ *or* $5 + (-2) = +3$

Answer: **+3**

Add a +5 arrow and a −2 arrow:

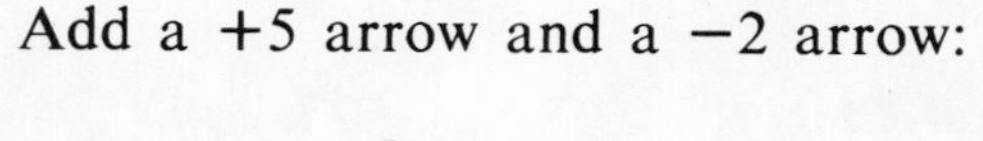

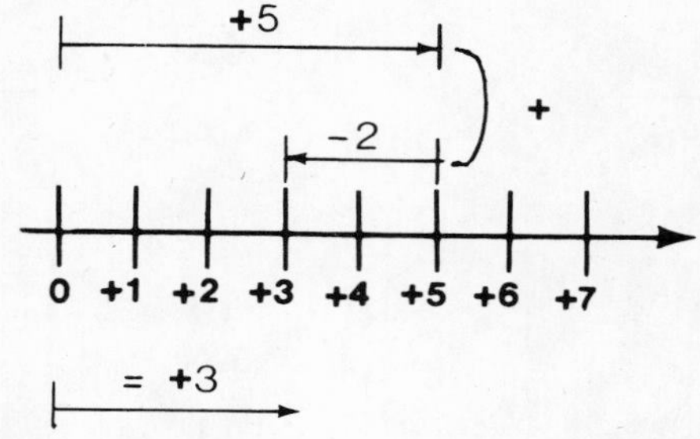

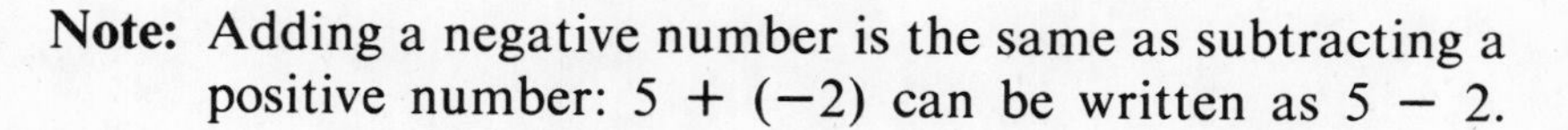

Note: Adding a negative number is the same as subtracting a positive number: $5 + (-2)$ can be written as $5 - 2$.

EXAMPLE 2. Add +3 and −7.

Step 1. Find the difference between 7 and 3.
$7 - 3 = 4$

Step 2. Give the (4) the sign (−) of the larger number (7).

Add: $\begin{array}{r} -7 \\ \underline{+3} \end{array}$ *or* $-7 + (+3) = -4$

Answer: **−4**

Add a +3 arrow and a −7 arrow:

EXAMPLE 3. Add $-4\frac{1}{2}$ and $+4\frac{1}{2}$.

Step 1. Find the difference between $4\frac{1}{2}$ and $4\frac{1}{2}$.
$4\frac{1}{2} - 4\frac{1}{2} = 0$

Answer: **0**

Add a $-4\frac{1}{2}$ arrow and a $+4\frac{1}{2}$ arrow:

= 0

Two numbers that differ only by sign are called *opposites*.
The sum of opposites is always 0.

Using Rule 2, add the following signed numbers.

Examples						
+9	**1.**	8	7	9	$5\frac{7}{9}$	6.5
−5		−6	−2	−7	−3	−4
4						
4	**2.**	6	1	3	8	7
−7		−9	−4	−6	$-9\frac{1}{4}$	−8.3
−3						
−9	**3.**	−8	−7	−12	$-10\frac{1}{2}$	−14.35
6		3	5	9	$8\frac{1}{4}$	12.20
−3						
−5	**4.**	−7	−2	−1	$-5\frac{1}{8}$	−11.25
+13		9	11	10	$8\frac{7}{8}$	14.75
8						
8	**5.**	−5	3	−4	$-8\frac{1}{2}$	12.67
−8		5	−3	4	$8\frac{1}{2}$	−12.67
0						

Examples		
8 + (−5) = 3	**6.** 9 + (−7) =	12 + (−5) =
7 + (−12) = −5	**7.** 1 + (−6) =	14 + (−24) =
−8 + (+6) = −2	**8.** −9 + (+7) =	−11 + 2 =
−7 + (+14) = 7	**9.** −4 + (+21) =	−9 + 13 =
−6 + (+6) = 0	**10.** −9 + (+9) =	14 + (−14) =

Using Rules 1 and 2, add the following signed numbers.

1.	$+9$ -8	-7 $+5$	-5 $+8$	-8 -3	$+6$ $-9\frac{1}{7}$	$+13.5$ $+5.5$
2.	-15 $+9$	-13 -12	21 -32	18 -9	$-14\frac{5}{6}$ $14\frac{5}{6}$	$+12$ -21.50
3.	17 -17	-8 -9	13 -8	$+15$ -21	-6 6	9.80 4
4.	34 53	-76 -54	-29 64	63 -19	$71\frac{2}{3}$ -35	-84 96.41
5.	$+3$ $+8$ $+7$	-4 -8 -7	-12 -32 -18	13 9 12	-32 -17 -9	-54 -39 -64
6.	21 53 38 15	$+31$ $+35$ $+73$ $+27$	-16 -19 -31 -24	-42 -38 -67 -53	75 52 53 48	-153 -214 -98 -79

7. $(-9) + (-6) =$ $\quad$ $-12 + (-3) =$ $\quad$ $8 + (+11) =$

8. $-21 + 9 =$ $\quad$ $9 + (-12) =$ $\quad$ $12 + 14 =$

9. $13 + 9 + 8 =$ $\quad$ $-3 + (-8) + (-7) =$ $\quad$ $23 + 31 + 29 =$

10. $-21 + (-31) + (-14) =$ $\quad$ $17 + 22 + 81 =$ $\quad$ $-35 + (-37) + (-28) =$

RULE 3: **To add several signed numbers at one time, combine the positive numbers and the negative numbers separately, and then add the positive and negative totals.**

EXAMPLE: Find the sum of +4, −9, −4, +8, +7, and −11.

Step 1. Combine the positive numbers and the negative numbers separately.

Positive Numbers	Negative Numbers
+4	−9
+8	−4
+7	−11
+19	−24

Step 2. Add the positive and negative totals by finding the difference between the two numbers. Give the answer the sign of the larger number.

Negative total	−24
Positive total	+19
Answer:	**−5**

Find the sum of each group of numbers.

1. +12, +9, −24, −8, +10, −7

2. −17, +13, −20, +9, +7, −4

3. −14, −5, +8, +7, +3, −6

4. +24, −10, −7, +8, −4, +5

5. +2.50, −4.75, +3.25, −5

6. −3.75, +2.50, +.75, −7

7. $+4\frac{1}{2}$, -3, $-1\frac{3}{4}$, $+2\frac{3}{4}$

8. $-10\frac{1}{2}$, $+8\frac{3}{4}$, -5, $+2\frac{1}{2}$

9. (+7) + (−8) + (+5) + (−9) =

10. (−8) + (−6) + (+7) + (+12) =

SUBTRACTING SIGNED NUMBERS

RULE: To subtract signed numbers, change the sign of the number being subtracted, and then follow the same steps that you used in adding signed numbers.

EXAMPLE 1. Subtract +5 from +8.

Step 1. Write $\begin{array}{r} +8 \\ \underline{+5} \end{array}$ *or* $+8 - (+5) =$

Step 2. Change the sign of the +5 to −5 and follow the rules for adding signed numbers.

Add: $\begin{array}{r} +8 \\ \underline{-5} \end{array}$ *or* $+8 + (-5) = \mathbf{+3}$

Answer: **+3**

> ***Note on Example 1***
> In algebra, subtracting a positive number is the same as adding a negative number.
> This is a new way to look at the operation of subtraction.

EXAMPLE 2. Subtract +12 from +3.

Step 1. Write $\begin{array}{r} +3 \\ \underline{+12} \end{array}$ *or* $+3 - (+12) =$

Step 2. Change the sign of the +12 to −12 and follow the rules for adding signed numbers.

Add: $\begin{array}{r} +3 \\ \underline{-12} \end{array}$ *or* $+3 + (-12) = \mathbf{-9}$

Answer: **−9**

> ***Note on Example 2***
> The subtraction rule allows us to subtract larger numbers from smaller numbers.
> This was not possible before studying signed numbers.

EXAMPLE 3. Subtract −4 from +7.

Step 1. Write $\begin{array}{r} +7 \\ \underline{-4} \end{array}$ *or* $+7 - (-4) =$

Step 2. Change the sign of the −4 to +4 and follow the rules for adding signed numbers.

Add: $\begin{array}{r} +7 \\ \underline{+4} \end{array}$ *or* $+7 + (+4) = \mathbf{+11}$

Answer: **+11**

> ***Note on Example 3***
> Subtracting signed numbers can be thought of as finding the distance between the two points on a number line. This is useful in seeing why the answer is larger than either the −4 or the +7.

−4 7 − (−4) = 11 +7

-5 -4 -3 -2 -1 0 +1 +2 +3 +4 +5 +6 +7 +8

Count the units between −4 and +7. Do you get 11?

Subtract the following signed numbers. The worked example at the beginning of each row reminds you to change the sign of the number being subtracted and then to follow the rules for adding signed numbers.

Examples

$$\begin{array}{r} +9 \\ \underline{+6} \end{array} \quad \text{add} \begin{bmatrix} +9 \\ \underline{-6} \\ 3 \end{bmatrix}$$

1. $\begin{array}{r} +7 \\ \underline{+5} \end{array}$ $\begin{array}{r} +8 \\ \underline{+7} \end{array}$ $\begin{array}{r} 12 \\ \underline{12} \end{array}$ $\begin{array}{r} 19 \\ \underline{14} \end{array}$ $\begin{array}{r} 32 \\ \underline{19} \end{array}$

$$\begin{array}{r} 5 \\ \underline{8} \end{array} \quad \text{add} \begin{bmatrix} 5 \\ \underline{-8} \\ -3 \end{bmatrix}$$

2. $\begin{array}{r} 7 \\ \underline{9} \end{array}$ $\begin{array}{r} 6 \\ \underline{7} \end{array}$ $\begin{array}{r} 9 \\ \underline{12} \end{array}$ $\begin{array}{r} 8 \\ \underline{13} \end{array}$ $\begin{array}{r} 11 \\ \underline{23} \end{array}$

$5 - (+9) =$
$5 + (-9) = -4$

$-2 - (-4) =$
$-2 + (+4) = 2$

$-8 - (-5) =$
$-8 + (+5) = -3$

3. $8 - (+11) =$ $11 - (6) =$

4. $-2 - (-5) =$ $-9 - (-15) =$

5. $-7 - (-6) =$ $-14 - (-5) =$

Here is some more practice subtracting signed numbers. Remember to change the sign of the number being subtracted.

6. $\begin{array}{r} -14 \\ \underline{-20} \end{array}$ $\begin{array}{r} 8 \\ \underline{-3} \end{array}$ $\begin{array}{r} -12 \\ \underline{7} \end{array}$ $\begin{array}{r} 13 \\ \underline{17} \end{array}$ $\begin{array}{r} 6 \\ \underline{-7} \end{array}$ $\begin{array}{r} 8 \\ \underline{5} \end{array}$

7. $\begin{array}{r} 9 \\ \underline{-3} \end{array}$ $\begin{array}{r} 11 \\ \underline{-19} \end{array}$ $\begin{array}{r} -24 \\ \underline{7} \end{array}$ $\begin{array}{r} -8 \\ \underline{12} \end{array}$ $\begin{array}{r} -19 \\ \underline{-26} \end{array}$ $\begin{array}{r} 20 \\ \underline{14} \end{array}$

8. $\begin{array}{r} 6 \\ \underline{-8} \end{array}$ $\begin{array}{r} -9 \\ \underline{-5} \end{array}$ $\begin{array}{r} -1 \\ \underline{10} \end{array}$ $\begin{array}{r} 15 \\ \underline{15} \end{array}$ $\begin{array}{r} 10 \\ \underline{-4} \end{array}$ $\begin{array}{r} 2 \\ \underline{3} \end{array}$

9. $(-5) - (-6) =$ $9 - (-7) =$ $-13 - (17) =$

10. $15 - (-23) =$ $21 - (9) =$ $-17 - (+17) =$

MULTIPLYING SIGNED NUMBERS

In arithmetic, multiplication is indicated by the times sign, "×". In algebra, multiplication is indicated by a dot "·" or by parentheses ( ). Study the following examples illustrating the multiplication of signed numbers.

$4 \cdot 5$ means "positive 4 times positive 5"
$-3 \cdot 6$ means "negative 3 times positive 6"
$4\,(+7)$ means "positive 4 times positive 7"
$(-9)\,(-8)$ means "negative 9 times negative 8"

Notice that parentheses may be placed around one or both signed numbers.

To multiply two signed numbers, follow these two rules:

RULE 1: If the signs of the numbers are alike, multiply the numbers and give the answer a positive sign.

RULE 2: If the signs of the numbers are different, multiply the numbers and give the answer a negative sign.

EXAMPLE 1. Multiply +4 and +3.
Step 1. Multiply the numbers.
Step 2. Give the answer a sign. Since the signs are alike, the sign of the answer is positive.

Multiply: $\begin{array}{r} +4 \\ \underline{+3} \\ \mathbf{+12} \end{array}$ *or* $+4\,(+3) = \mathbf{+12}$

EXAMPLE 2. Find the product of −5 and −6.
Step 1. Multiply the numbers.
Step 2. Give the answer a sign. Like signs make the sign of the answer positive.

Multiply: $\begin{array}{r} -5 \\ \underline{-6} \\ \mathbf{+30} \end{array}$ *or* $(-5)(-6) = \mathbf{+30}$

EXAMPLE 3. What is −7 times +2?
Step 1. Multiply the numbers.
Step 2. Give the answer a sign. Unlike signs make the sign of the answer negative.

Multiply: $\begin{array}{r} -7 \\ \underline{+2} \\ \mathbf{-14} \end{array}$ *or* $-7 \cdot 2 = \mathbf{-14}$

Multiply the following signed numbers. Remember: Like signs give positive answers. Unlike signs give negative answers.

Examples

$+7$	**1.**	$+9$	6	$+10$	12	$+14$
$\underline{+4}$		$\underline{+3}$	$\underline{7}$	$\underline{+9}$	$\underline{11}$	$\underline{+10}$
$+28$						
-9	**2.**	-5	-7	-8	-13	-12
$\underline{-2}$		$\underline{-3}$	$\underline{-4}$	$\underline{-9}$	$\underline{-5}$	$\underline{-14}$
$+18$						
$+5$	**3.**	-7	-3	$+9$	11	-23
$\underline{-4}$		$\underline{+6}$	$\underline{8}$	$\underline{-4}$	$\underline{-7}$	$\underline{+16}$
-20						

$(+6)(+4) = +24$ **4.** $5 \cdot 3 =$ $+8\ (+7) =$ $(9)(5) =$

$-7\ (-9) = +63$ **5.** $(-8)(-7) =$ $6 \cdot 8 =$ $(-4)(-8) =$

$(+4)(-7) = -28$ **6.** $(-2)(+5) =$ $(-3)(+8) =$ $7\ (-7) =$

Multiply the following signed numbers.

7.	$+8$	-5	7	$+3$	-19	$+24$
	$\underline{-6}$	$\underline{-9}$	$\underline{5}$	$\underline{+6}$	$\underline{13}$	$\underline{-18}$
8.	-9	-4	6	$+28$	13	-42
	$\underline{9}$	$\underline{+2}$	$\underline{-7}$	$\underline{-14}$	$\underline{11}$	$\underline{-31}$
9.	$+13$	-17	32	-29	-37	$+68$
	$\underline{-10}$	$\underline{-9}$	$\underline{-27}$	$\underline{+17}$	$\underline{-28}$	$\underline{-45}$

10. $(-7)(-12) =$ $9\ (-8) =$ $(-4)(+13) =$ $-13 \cdot 12 =$

11. $-15 \cdot 17 =$ $(21)(18) =$ $(+8)(-3) =$ $24 \cdot 25 =$

DIVIDING SIGNED NUMBERS

In algebra, you indicate division the same way that you do in arithmetic—with a fraction bar or with a division sign (÷). To avoid confusion, parentheses are sometimes placed around signed numbers when the division sign is used.

$-15 \div 5$ means "negative 15 divided by positive 5"
$(+12) \div (-6)$ means "positive 12 divided by negative 6"
$(-18) \div (-3)$ means "negative 18 divided by negative 3"

The rules for dividing signed numbers are similar to the rules for mutiplying signed numbers.

To divide signed numbers, follow these two rules:

RULE 1: If the signs of the numbers are alike, divide the numbers and give the answer a positive sign.

RULE 2: If the signs of the numbers are different, divide the numbers and give the answers a negative sign.

EXAMPLE 1. Divide +20 by +4. $\frac{+20}{+4} = \mathbf{+5}$ *or* $+20 \div (+4) = \mathbf{+5}$
Step 1. Divide the numbers.
Step 2. Give the answer a sign. Since the signs are alike, the sign of the answer is positive.

EXAMPLE 2. Divide −36 by −9. $\frac{-36}{-9} = \mathbf{+4}$ *or* $(-36) \div (-9) = \mathbf{+4}$
Step 1. Divide the numbers.
Step 2. Give the answer a sign. Since the signs are alike, the answer is positive.

EXAMPLE 3. Divide −48 by +8. $\frac{-48}{+8} = \mathbf{-6}$ *or* $(-48) \div (+8) = \mathbf{-6}$
Step 1. Divide the numbers
Step 2. Give the answer a sign. Since the signs are unlike, the answer is negative.

EXAMPLE 4. Divide +3 by −4. $\frac{+3}{-4} = \mathbf{-\frac{3}{4}}$ *or* $+3 \div (-4) = \mathbf{-\frac{3}{4}}$
Step 1. Divide the numbers.
Step 2. Give the answer a sign. Since the signs are unlike, the answer is negative.

Notice that the sign is placed in front of the fraction.

Here are two more examples of fractions:

$$\frac{+2}{-4} = -\frac{2}{4} = -\frac{1}{2} \qquad \frac{-3}{-9} = +\frac{3}{9} = +\frac{1}{3} \quad or \quad \frac{1}{3}$$

Notice that the fractions have been reduced to lowest terms.

Divide the following signed numbers. *Remember:* Like signs give positive answers. Unlike signs give negative answers.

Examples

$\frac{+8}{+4} = +2$

$\frac{-15}{-3} = +5$

$\frac{+16}{-4} = -4$

1. $\frac{+12}{+3} =$ $\qquad \frac{3}{9} =$ $\qquad \frac{+24}{+6} =$ $\qquad \frac{18}{30} =$

2. $\frac{-20}{-4} =$ $\qquad \frac{-42}{-14} =$ $\qquad \frac{-9}{-12} =$ $\qquad \frac{-12}{-15} =$

3. $\frac{-4}{12} =$ $\qquad \frac{-14}{+7} =$ $\qquad \frac{32}{-8} =$ $\qquad \frac{11}{-33} =$

Examples

$(+18) \div (+9) = +2$

$(-32) \div (-4) = +8$

$-9 \div 12 = -\frac{9}{12} = -\frac{3}{4}$

4. $28 \div 7 =$ $\qquad (+14) \div (+20) =$

5. $(-63) \div (-9) =$ $\qquad (-9) \div (-15) =$

6. $(+26) \div (-13) =$ $\qquad (-3) \div (27) =$

Divide the following signed numbers.

7. $\frac{+24}{-6} =$ $\qquad \frac{-40}{-10} =$ $\qquad \frac{8}{8} =$ $\qquad \frac{-14}{+2} =$ $\qquad \frac{9}{-15} =$

8. $\frac{+13}{-39} =$ $\qquad \frac{45}{-9} =$ $\qquad \frac{-15}{+5} =$ $\qquad \frac{8}{-12} =$ $\qquad \frac{-38}{+24} =$

9. $\frac{+35}{-7} =$ $\qquad \frac{-8}{14} =$ $\qquad \frac{-35}{+7} =$ $\qquad \frac{36}{12} =$ $\qquad \frac{-56}{-7} =$

10. $(+81) \div (-9) =$ $\qquad (+72) \div (+8) =$ $\qquad (-16) \div (-48) =$

ADDITION AND SUBTRACTION OF SIGNED NUMBERS

As you saw on page 11, more than two numbers may be added together. In some cases, you may need to combine addition and subtraction of signed numbers, as in this expression:

$$10 - (+5) - (-2) + (+4) - (+7) =$$

To simplify (solve) this expression, follow these steps:

Step 1. Change the sign of each number being subtracted. Change each subtraction sign to an addition sign.

Step 2. Combine the positive numbers and combine the negative numbers separately.

Step 3. Add the positive and negative totals by finding the difference between the two numbers. Give your answer the sign of the larger number.

EXAMPLE. Simplify: $10 - (+5) - (-2) + (+4) - (+7)$

Step 1. Change the sign of each number being *subtracted.*

Change each subtraction sign to an addition sign.

Write: $10 + (-5) + (+2) + (+4) + (-7)$

Note: Don't change the sign of the number being added (+4).

Step 2. Combine the positive numbers. Combine the negative numbers.

Positive Numbers	Negative Numbers
10	−5
2	−7
4	
16	−12

Step 3. Find the difference between the two numbers.
Give the answer the sign of the larger number.

Positive total	16
Negative total	−12
	4

Answer: 4

Simplify each of the following.

1. $15 + (-6) - (-7) + (+5) =$

2. $25 - (+5) - (-9) - (+14) =$

3. $-19 - (-9) - (+11) + (+3) - (-8) =$

4. $17 + (-2) + (-8) - (-10) - (-12) =$

MULTIPLICATION OF MORE THAN 2 SIGNED NUMBERS

On pages 14 and 15, you learned how to multiply two signed numbers. To multiply several signed numbers, multiply the numbers two at a time, following the rules that you learned earlier.

EXAMPLE 1. Multiply $(-3)(+4)(-2)$

Step 1. Multiply (-3) by $(+4)$

$(-3)(+4) = -12$

Step 2. Multiply (-12) by (-2)

$(-12)(-2) = \mathbf{+24}$

Answer: 24

EXAMPLE 2. Multiply $(+5)(-6)(+3)$

Step 1. Multiply $(+5)$ by (-6)

$(+5)(-6) = -30$

Step 2. Multiply (-30) by $(+3)$

$(-30)(+3) = \mathbf{-90}$

Answer: −90

A quick way to check the sign of the answer is to count how many negative numbers are being multiplied:

1. An even number of negative signs gives a positive answer.
2. An odd number of negative signs gives a negative answer.

Example 1 has two negative signs; two is an even number, which makes the answer positive. Example 2 has one negative sign; one is an odd number, which makes the answer negative.

EXAMPLE 3. Multiply $(-2)(-5)(+2)(-4)$

Step 1. $(-2)(-5) = +10$

Step 2. $(+10)(+2) = +20$

Step 3. $(+20)(-4) = \mathbf{-80}$

Answer: −80

Shortcut to Example 3

Multiply the numbers together.

$2 \cdot 5 = 10;\ 10 \cdot 2 = 20;\ 20 \cdot 4 = 80$

Answer: −80 The answer is negative since three (odd) negative numbers are being multiplied.

Multiply the following signed numbers.

1. $(-5)(+4)(-7) =$

2. $(-6)(-9)(-3) =$

3. $(+10)(+2)(-3) =$

4. $(3)(7)(+8) =$

5. $(+3)(-9)(+4) =$

6. $(-8)(-6)(+4) =$

7. $(2)(5)(-6)(-3) =$

8. $(-4)(-5)(-6)(-1) =$

9. $(-4)(3)(-7)(-2) =$

SIGNED NUMBERS: APPLYING YOUR SKILLS

Signed numbers are often used to represent opposite quantities. For example, positive numbers represent temperatures above 0°, distances above sea level, and payments and credits. Negative numbers represent temperatures below 0°, distances below sea level, and charges or withdrawals.

Study the following examples before beginning the word problems on the next page.

EXAMPLE 1. During the day, the temperature rose from −5° F to +17° F. How many degrees did the temperature rise?

The change in temperature is found by subtracting the earlier temperature (−5°)from the later temperature (+17°).

Change in temperature $= +17^\circ - (-5^\circ)$

$= +17^\circ + 5^\circ = \mathbf{+22^\circ}$

Answer: The temperature rose **22°**.

EXAMPLE 2. On June 1, Mary owed $195.00 on her charge card. During June, she made the following charges and payments, and she returned two items for credit. What was her new balance at the end of June?

6/3	Charge	$34.00	6/2	Credit	$64.00
6/9	Charge	17.75	6/15	Payment	50.00
6/26	Charge	24.25	6/21	Credit	14.50
6/27	Charge	13.99	6/30	Payment	45.00
Finance charge $1.50					

Step 1. Add the beginning balance, charges, and finance charge as negative numbers because these are what she owes. Add credits and payments as positive numbers because these are amounts that Mary has returned or paid back to the store.

Negative Total		Positive Total	
−195.00	balance owed	+64.00	credit
−34.00	charge	+50.00	payment
−17.75	charge	+14.50	credit
−24.25	charge	+45.00	payment
−13.99	charge	+173.50	
−1.50	finance charge		
−286.49			

Step 2. Add the negative and positive totals. The negative sum is the new balance on her charge card.

−286.49
+173.50

Answer: **−112.99** The new balance is **$112.99**. This is what she owes.

Solve the following problems.

1. During the evening, the temperature was $-6°F$. By morning, the temperature was $+7°F$. How many degrees did the temperature rise overnight?

____

2. The lowest recorded temperature in the U.S. occurred on January 23, 1971 in Prospect Creek, Alaska. On that day, the temperature fell to $-79°F$. How many degrees below freezing did the temperature fall? (Freezing = 32°F)

____

3. Mount McKinley, the highest mountain in North America, is 20,320 feet above sea level. Death Valley, the lowest point in North America, is 280 feet below sea level. Find the difference in height between Mount McKinley and Death Valley.

____

4. Mount Whitney, the highest mountain in the U.S., excluding Alaska, is 14,495 feet high. What is the difference in height between Mount Whitney and Death Valley? (See problem 3.)

____

5. On September 1, John owed $345.50 on his charge card. During September, he made the following charges and payments, and he returned two items for credit. Find his new balance at the end of the month.

9/3	Charge	$ 45.00	9/2	Payment	$65.00
9/7	Charge	76.49	9/4	Credit	14.00
9/24	Charge	160.75	9/20	Payment	34.79
9/27	Charge	28.00	9/25	Credit	16.40
Finance Charge		2.00			

____

SIGNED NUMBERS SKILLS INVENTORY

1. Write each number below as a signed number using a + sign or a − sign.

positive 13 =

positive $2\frac{3}{4}$ =

negative 26 =

negative $\frac{7}{8}$ =

2. Locate these points on the number line below: -6, $-3\frac{1}{2}$, $-\frac{1}{2}$, $+\frac{3}{4}$, $+2\frac{1}{2}$, $+7$.

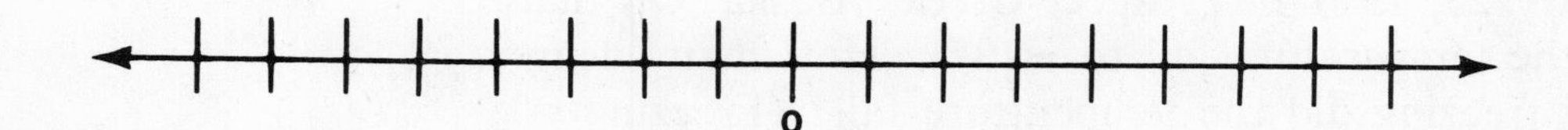

3. Add: $\begin{array}{r} -17 \\ \underline{-12} \end{array}$

4. $-21 + (-9) =$

5. Add: $\begin{array}{r} -7 \\ \underline{+13} \end{array}$

6. $+8 + (-15) =$

7. What is the opposite of $+9\frac{1}{3}$?

8. What is the sum of $-13\frac{2}{3}$ and $+13\frac{2}{3}$?

9. Find the total: $\begin{array}{r} +17 \\ -12 \\ +8 \\ +7 \\ \underline{-10} \end{array}$

10. Subtract: $\begin{array}{r} -43 \\ \underline{-18} \end{array}$

11. $(+27) - (-19) =$

12. $-17 - (+9) =$

13. $(-7) - (-10) + (3) - (6) =$

14. Multiply: $\begin{array}{r} -17 \\ \underline{-8} \end{array}$

15. $(-13)(-7) =$

16. Multiply: $\begin{array}{r} +29 \\ \underline{-4} \end{array}$

17. $-4(+7)(+2) =$

18. Divide: $\frac{-18}{-6} =$

19. $(-15) \div (-25) =$

20. $\frac{-36}{+9} =$

21. $(+24) \div (-6) =$

22. At 10:00 a.m., the temperature was $-6°$F. By 3:00 p.m., the temperature had risen to $+17°$F. How many degrees did the temperature rise?

23. On March 1st, Jim had a balance of $245 in his checking account. During the month of March, Jim wrote checks and made deposits as follows:

3/12	Check $ 25.50	3/15	Deposit $ 50.00
3/17	Check $ 15.75	3/30	Deposit $450.00
3/20	Check $ 55.00		
3/30	Check $345.00		

At the end of March, what was the new balance in Jim's checking account?

SIGNED NUMBERS INVENTORY CHART

Circle the number of any problem that you missed and be sure to review the appropriate practice page. A passing score is 20 correct answers.

Problem Number	Skill Area	Practice Page	Problem Number	Skill Area	Practice Page
1	introduction	4	13	combined addition and subtraction	18
2	number line	5	14	multiplication	14
3	addition	6	15	multiplication	14
4	addition	6	16	multiplication	14
5	addition	8	17	multiplication	19
6	addition	8	18	division	16
7	opposites	8	19	division	16
8	opposites	8	20	division	16
9	addition	11	21	division	16
10	subtraction	12	22	applying skills	20
11	subtraction	12	23	applying skills	20
12	subtraction	12			

If you missed more than 3 questions, you should review the chapter.

POWERS AND ROOTS

WHAT IS A POWER?

A *power* is the product of a number multiplied by itself one or more times. Four to the third power means "4 times 4 times 4."

A power is commonly written as a base and an exponent:

$$(4)(4)(4) \text{ is written } 4^3$$

(in 4^3, the 3 is labeled "exponent" and the 4 is labeled "base")

The 4 is called the *base*. The base is the number being multiplied.

The 3 is called the *exponent*. The exponent tells how many times the base is written in the product.

Look at these examples:

Product	As a base and exponent	In Words
$3 \cdot 3$	3^2	three to the second power or three squared
$(-4)(-4)(-4)$	$(-4)^3$	negative four to the third power or negative four cubed
$(\frac{1}{2})(\frac{1}{2})(\frac{1}{2})(\frac{1}{2})$	$(\frac{1}{2})^4$	one-half to the fourth power

Note: A number raised to the second power is often called squared and raised to the third power is often called cubed. These are the only two powers that have special names.

The *value* of a power is found by multiplication.

EXAMPLE 1. Find the value of 4^4.

Step 1. Write out the terms for the base and exponent.

Step 2. Multiply the first two terms together: $(4)(4) = 16$

Step 3. Multiply the answer in Step 2 times the next number. $(16)(4) = 64$

Step 4. Continue multiplying until you have used all the terms. $(64)(4)$

Answer: $= \mathbf{256}$

$$\begin{aligned} 4^4 &= (4)(4)(4)(4) \\ &= (16)(4)(4) \\ &= (64)(4) \\ &= \mathbf{256} \end{aligned}$$

EXAMPLE 2. Find the value of $(-\frac{2}{3})^3$

Step 1. Write out your terms. $(-\frac{2}{3})^3 = (-\frac{2}{3})(-\frac{2}{3})(-\frac{2}{3})$

Step 2. Multiply $(-\frac{2}{3})(-\frac{2}{3}) = +\frac{4}{9}$ $= (+\frac{4}{9})(-\frac{2}{3})$

Step 3. Multiply $(+\frac{4}{9})(-\frac{2}{3})$ $= -\frac{8}{27}$

Answer: $-\frac{8}{27}$

Be sure that the sign of your answer is correct.

Complete the following chart.

	Product	As a base and exponent	In Words
1.	$(-1)(-1)$	$(-1)^2$	negative one to the second power or negative one squared
2.	$(+5)(+5)$		
3.	$(+\frac{1}{2})(+\frac{1}{2})$		
4.	$6 \cdot 6$		
5.	$7 \cdot 7 \cdot 7$	7^3	seven to the third power or seven cubed
6.	$(-3)(-3)(-3)$		
7.	$(+6)(+6)(+6)$		
8.	$(-\frac{2}{3})(-\frac{2}{3})(-\frac{2}{3})$		
9.	$(+2)(+2)(+2)(+2)$	$(+2)^4$	positive two to the fourth power
10.	$5 \cdot 5 \cdot 5 \cdot 5$		
11.	$(-4)(-4)(-4)(-4)$		
12.	$(+\frac{1}{4})(+\frac{1}{4})(+\frac{1}{4})(+\frac{1}{4})$		

Find the value of each power below.

13. $5^3 =$ **14.** $(-2)^2 =$ **15.** $(-6)^3 =$ **16.** $(+9)^2 =$

17. $(-5)^4 =$ **18.** $(+\frac{2}{3})^2 =$ **19.** $(-10)^3 =$ **20.** $(-3)^3 =$

21. $5^4 =$ **22.** $(-\frac{4}{5})^3 =$ **23.** $(+5)^2 =$ **24.** $(-2)^3 =$

There are two special cases that you should learn:

(1) Any number to the first power is that number.
EXAMPLE: $8^1 = 8$
An exponent 1 means the base (8) is written only once and is not multiplied.

(2) Any number to the zero power is one.
EXAMPLE: $6^0 = 1$
This is the hardest rule to remember. A zero exponent means "a number divided by itself."

To simplify (solve) an expression containing more than one power, find the value for each power and then follow the rules for adding and subtracting signed numbers.

EXAMPLE: Simplify the expression $(-4)^3 - (3)^4 + (7)^0$

Step 1. Find the value of each power.
$(-4)^3 = -64 \quad (3)^4 = 81 \quad (7)^0 = 1$

Step 2. Substitute the values and follow the rules for adding and subtracting signed numbers.

$$\begin{aligned}(-4)^3 - (3)^4 + (7)^0 &= -64 - (81) + 1 \\ &= \underbrace{-64 - 81} + 1 \\ &= -145 + 1\end{aligned}$$

Answer: $= \mathbf{-144}$

Simplify each expression below.

25. $8^2 - 3^2 =$

26. $9^1 - (3)^2 + (4)^4 =$

27. $7^2 + 3^3 - 4^1 =$

28. $5^0 + (-3)^2 + 4^1 =$

29. $4^3 - (3)^2 + (5)^0 =$

30. $(\frac{1}{2})^2 + (\frac{1}{2})^3 =$

31. $(-4)^3 - (-3)^3 + 2^4 =$

32. $(+\frac{3}{4})^2 - (-\frac{1}{2})^2 + (+\frac{1}{2})^1 =$

MULTIPLICATION AND DIVISION OF POWERS

You may need to find the value of a product or a quotient of powers. In either case, first find the value of each power and then multiply or divide as indicated.

EXAMPLE 1. Find the value of the product $3^2 \cdot 2^4$.

Step 1. Find the value of each power.
$3^2 = 9$ and $2^4 = 16$

Step 2. Multiply the values found in Step 1.
$3^2 \cdot 2^4 = 9 \cdot 16 = 144$

Answer: **144**

EXAMPLE 2. Find the value of the quotient $\frac{4^3}{2^2}$.

Step 1. Find the value of each power.
$4^3 = 64$ and $2^2 = 4$

Step 2. Divide 64 by 4.
$\frac{64}{4} = 16$

Answer: **16**

Find the value of each product or quotient below.

Examples

$2^3 \cdot 4^2 = 8 \cdot 16$
$= 128$

$(\frac{1}{2})^3 \cdot 5^2 = (\frac{1}{8}) \cdot 25$
$= 3\frac{1}{8}$

$\frac{5^3}{7^2} = \frac{125}{49}$
$= 2\frac{27}{49}$

1. $7^2 \cdot 3^3 =$

2. $5^3 \cdot 3^2 =$

3. $-2^3 \cdot 4^2 =$

4. $(\frac{1}{5})^2 \cdot 5^3 =$

5. $2^2(\frac{3}{4})^3 =$

6. $(\frac{1}{4})^2(\frac{2}{5})^3 =$

7. $\frac{3^4}{7^2} =$

8. $\frac{-4^3}{2^4} =$

9. $\frac{6^2}{3^4} =$

To solve more complicated problems, first find the value of each power and then use cancellation when possible to simplify the multiplication and division.

Example

$\frac{8^2 \cdot 4^1}{2^3 \cdot 3^2} = \frac{{}^{8}\cancel{64} \cdot 4}{\cancel{8}_{1} \cdot 9}$
$= \frac{8 \cdot 4}{9} = \frac{32}{9}$
$= 3\frac{5}{9}$

10. $\frac{7^2 \cdot 2^3}{4^2 \cdot 3^2} =$

11. $\frac{4^2 \cdot 5^2}{-6^3 \cdot 2^4} =$

12. $\frac{3^2 \cdot 5^1}{5^2 \cdot 3^4} =$

WHAT IS A SQUARE ROOT?

Another important skill to learn is finding a *square root*. To find the square root of a number ask yourself, "What number times itself equals this number?"

For example, since $5^2 = 25$, 5 is the square root of 25.
The square root symbol is $\sqrt{}$. Thus, $5 = \sqrt{25}$.

EXAMPLE 1. Seven is the square root of what number?
Since $7^2 = 49$, 7 is the square root of 49.

Answer: $7 = \sqrt{\mathbf{49}}$

EXAMPLE 2. Find $\sqrt{36}$
Since $6^2 = 36$, the square root of 36 is 6.

Answer: $\sqrt{36} = \mathbf{6}$

Numbers that have whole number square roots are called *perfect squares*. Perfect squares are easily found by "squaring" whole numbers. The first fifteen perfect squares are shown on the table below. They are important to learn.

Table of Perfect Squares

$1^2 = 1$	$6^2 = 36$	$11^2 = 121$
$2^2 = 4$	$7^2 = 49$	$12^2 = 144$
$3^2 = 9$	$8^2 = 64$	$13^2 = 169$
$4^2 = 16$	$9^2 = 81$	$14^2 = 196$
$5^2 = 25$	$10^2 = 100$	$15^2 = 225$

As you already know, 0 times 0 is 0. Therefore, $0^2 = 0$.

To find the square root of a perfect square from the table, find the perfect square in the right-hand column. Read the whole number square root in the left-hand column. For example, the square root of 121 is 11.

Write each sentence in symbols.

1. Three is the square root of nine. $\mathbf{3 = \sqrt{9}}$
2. Seven is the square root of forty-nine. ______________
3. Ten is the square root of one hundred. ______________
4. Twelve is the square root of one hundred forty-four. ______________
5. Fifteen is the square root of two hundred twenty-five. ______________

From the table on the previous page, find the square root of each perfect square below.

1. $\sqrt{169}$ = $\sqrt{25}$ = $\sqrt{81}$ =

2. $\sqrt{4}$ = $\sqrt{121}$ = $\sqrt{36}$ =

3. $\sqrt{49}$ = $\sqrt{1}$ = $\sqrt{225}$ =

4. $\sqrt{144}$ = $\sqrt{64}$ = $\sqrt{9}$ =

5. $\sqrt{100}$ = $\sqrt{16}$ = $\sqrt{196}$ =

The square root of a fraction is equal to the square root of the numerator over the square root of the denominator.

EXAMPLE: Find $\sqrt{\frac{36}{49}}$

Step 1. Rewrite as the square root of the numerator over the square root of the denominator. $\sqrt{\frac{36}{49}} = \frac{\sqrt{36}}{\sqrt{49}}$

Step 2. Find the square roots of both the numerator and denominator. $= \frac{6}{7}$

Answer: $\frac{6}{7}$

Check: $(\frac{6}{7})^2 = \frac{36}{49}$

Find the square root of each fraction below. Check each answer.

6. $\sqrt{\frac{4}{9}}$ = $\sqrt{\frac{9}{64}}$ = $\sqrt{\frac{25}{36}}$ =

7. $\sqrt{\frac{1}{4}}$ = $\sqrt{\frac{49}{81}}$ = $\sqrt{\frac{121}{144}}$ =

8. $\sqrt{\frac{81}{16}}$ = $\sqrt{\frac{1}{196}}$ = $\sqrt{\frac{100}{121}}$ =

FINDING AN APPROXIMATE SQUARE ROOT

A number that is not a perfect square does not have a whole number square root. For example, the square root of 30 is between the whole numbers 5 and 6:

$\sqrt{30}$ is larger than 5, since $5^2 = 25$
$\sqrt{30}$ is smaller than 6, since $6^2 = 36$

To find the approximate square root of a number that is not a perfect square, follow the three steps of the Method of Averaging:

Step 1. Choose a number that is close to the correct square root.
Step 2. Divide this chosen number into the number you're trying to find the square root of.
Step 3. Average your choice from Step 1 with the answer from Step 2. The average of these two numbers is the approximate square root.

Note: The approximation sign "$\approx$" means "is approximately equal to."

EXAMPLE 1. Find the approximate square root of 30.

Step 1. Find the perfect square that is closest to the number 30. This perfect square is 25. Write 5, the exact square root of 25. This is approximately equal to the square root of 30.

$5 \approx \sqrt{30}$
since $5 = \sqrt{25}$

Step 2. Divide 5 into the original number, 30.
$30 \div 5 = 6$

$$\begin{array}{r} 6 \\ 5\overline{)30} \\ \underline{30} \end{array}$$

Step 3. Find the average of the two numbers (5 and 6) that you found in Steps 1 and 2. This average is the approximate square root of 30.

$5 + 6 = 11$

$$\begin{array}{r} 5\frac{1}{2} \\ 2\overline{)11} \\ \underline{10} \\ 1 \end{array}$$

Answer: $5\frac{1}{2} = \mathbf{5.5}$
Check: $(5.5)^2 = 30.25$

EXAMPLE 2. Find the approximate square root of 54.

Step 1. Choose $7 \approx \sqrt{54}$ since $7 = \sqrt{49}$

Step 2. Divide 54 by 7.
$54 \div 7 \approx 7.71$

$$\begin{array}{r} 7.71 \\ 7\overline{)54.00} \\ \underline{49} \\ 50 \\ \underline{49} \\ 10 \\ \underline{7} \\ 3 \end{array}$$

Step 3. Average the numbers 7 and 7.71.

$$\begin{array}{r} 7.71 \\ +\ \underline{7} \\ 14.71 \end{array} \qquad \begin{array}{r} 7.35\frac{1}{2} \\ 2\overline{)14.71} \\ \underline{14} \\ 7 \\ \underline{6} \\ 11 \\ \underline{10} \\ 1 \end{array}$$

$7.35\frac{1}{2}$ rounded off is **7.36**

Answer: $\mathbf{7.36} \approx \sqrt{54}$

Check: $(7.36)^2 = 54.1696$

Note: The average in Step 3 is carried out to two decimal places. Two decimal place accuracy gives a good approximation, as shown by checking the answer 7.36.

Use the Method of Averaging to find an approximate square root in each problem below. Check each answer.

1. $\sqrt{42} \approx$

2. $\sqrt{19} \approx$

3. $\sqrt{76} \approx$

4. $\sqrt{90} \approx$

5. $\sqrt{28} \approx$

6. $\sqrt{107} \approx$

POWERS AND ROOTS SKILLS INVENTORY

1. In the expression 3^4, ____ is called the base, and ____ is called the exponent.

Problems 2-7: Write each product as a base and an exponent.

2. $(-4)(-4)(-4)(-4)$ ______

3. $(+5)(+5)(+5)$ ______

4. $(-\frac{3}{5})(-\frac{3}{5})$ ______

5. $8 \cdot 8$ ______

6. $(+\frac{5}{6})(+\frac{5}{6})(+\frac{5}{6})$ ______

7. $9 \cdot 9 \cdot 9 \cdot 9$ ______

Problems 8-13: Find the value of each power.

8. $(+9)^2 =$

9. $(-5)^3 =$

10. $(-3)^4 =$

11. $(-\frac{2}{3})^3 =$

12. $(+4)^4 =$

13. $(-\frac{4}{5})^2 =$

14. What is the value of 3^0?

15. Find the value of 7^1.

16. Simplify $5^3 - (-3)^4 + 5^0$

17. Simplify $\dfrac{4^3 \cdot 2^2}{-2^3 \cdot 2^1}$

18. Six is the square root of what number?

19. $\sqrt{49} =$

20. $\sqrt{\frac{81}{121}} =$

21. Find an approximate square root of 56.

22. Find an approximate square root of 83.

POWERS AND ROOTS INVENTORY CHART

Circle the number of any problems that you missed and be sure to review the appropriate practice page. A passing score is 19 correct answers.

Problem Number	Skill Area	Practice Page
1	powers	24
2	powers	24
3	powers	24
4	powers	24
5	powers	24
6	powers	24
7	powers	24
8	powers	24
9	powers	24
10	powers	24
11	powers	24
12	powers	24
13	powers	25
14	powers	26
15	powers	26
16	powers	26
17	powers	27
18	roots	28
19	roots	28
20	roots	29
21	approximate square roots	30
22	approximate square roots	30

If you missed more than 3 questions, you should review the chapter.

ALGEBRAIC EXPRESSIONS

WRITING ALGEBRAIC EXPRESSIONS

In algebra, we often use letters to represent numbers. A letter that stands for a number is called a *variable* or an *unknown.*

A variable can be used to represent numbers in addition, subtraction, multiplication, or division problems. The symbols used in algebra are "+" for addition and "−" for subtraction. Multiplication is indicated by placing a number next to a variable; no multiplication sign is used. Division is indicated by placing a number or variable over the other.

An *algebraic expression* consists of two or more numbers or variables combined by one or more of the operations—addition, subtraction, multiplication, or division.

The following are examples of algebraic expressions:

Operation	Algebraic Expression	Word Expression
Addition	$x + 2$	x plus 2
Subtraction	$y - 3$	y minus 3
	or	
	$3 - y$	3 minus y
Multiplication	$4z$	4 times z
Division	$\frac{n}{8}$	n divided by 8
	or	
	$\frac{8}{n}$	8 over n

Write an algebraic expression for each word expression below.

1. A number x plus nine

2. Two times a number y

3. y plus negative 2

4. Five x

5. One subtracted from z

6. The product of 3 and x

7. r divided by 22

8. The sum of x plus negative 4

9. w minus 7

10. A number z subtract 12

11. A number x divided by 5

12. Negative x over 12

13. c minus 4

14. A number y minus negative nine

15. Three plus n

16. A number m decreased by 3

17. Thirteen divided by a number y **18.** r over negative 22

Many algebraic expressions contain more than one of the operations of addition, subtraction, multiplication, or division. Placing a number or variable outside of an expression in parentheses means that the whole expression is to be multiplied by the term on the outside.

For example, look at the difference in meaning between $3y+7$ and $3(y+7)$. If the number 2 were substituted for y in each expression, the following solutions would result:

$$3y + 7 = 3 \cdot 2 + 7 = 6 + 7 = \mathbf{13} \quad \textit{but} \quad 3(y + 7) = 3(2 + 7) = 3(9) = \mathbf{27}$$

The following examples show how parentheses can change the meaning of an algebraic expression.

Algebraic Expression	Word Expression
$3y - 7$	3 times y minus 7
$3(y - 7)$	3 times the quantity y minus 7
$-\frac{2}{3}x + 5$	Negative $\frac{2}{3}x$ plus 5
$-\frac{2}{3}(x + 5)$	Negative $\frac{2}{3}$ times the quantity x plus 5
$3x^2 + 2$	3 x-squared plus 2
$3(x^2 + 2)$	3 times the quantity x-squared plus 2

Match each algebraic expression with its equivalent word expression. Write the letter of the word expression on the line to the left of each algebraic expression.

	Algebraic Expressions	Equivalent Word Expressions
______ **19.**	$4x - 9$	**a.** z-squared minus 4
______ **20.**	$\frac{2}{5}(y + 6)$	**b.** Two-fifths y plus 6
______ **21.**	$-2a(a + 7)$	**c.** z times the quantity z minus 4
______ **22.**	$4(x - 9)$	**d.** Negative $\frac{3}{4}$ times the quantity x minus 4
______ **23.**	$\frac{2}{5}y + 6$	**e.** The sum of negative two times a-squared and 7
______ **24.**	$-2a^2 + 7$	**f.** Two-fifths times the quantity y plus 6
______ **25.**	$-\frac{3}{4}x - 4$	**g.** Negative two times a times the sum of a and 7
______ **26.**	$z^2 - 4$	**h.** Nine subtracted from the product of 4 times x
______ **27.**	$-\frac{3}{4}(x - 4)$	**i.** Four times the quantity x minus 9
______ **28.**	$z(z - 4)$	**j.** Negative three-fourths x take away 4

EVALUATING ALGEBRAIC EXPRESSIONS

You can find the value of an algebraic expression by substituting a number for a letter and doing the necessary arithmetic.

EXAMPLE 1. Find the value of $x + 2$ for $x=3$.

Step 1. Substitute the number for the letter. $x + 2 = (3) + 2$

Step 2. Do the arithmetic. $= \mathbf{5}$

Answer: **5**

EXAMPLE 2. Find the value of rt for $r=40$ and $t=3$.

Step 1. Substitute. $rt = (40)(3)$

Step 2. Solve. $= \mathbf{120}$

Answer: **120**

EXAMPLE 3. Find the value for $\frac{d}{r}$ for $d=200$ and $r=50$.

Step 1. Substitute $\frac{d}{r} = \frac{200}{50}$

Step 2. Solve. $= \mathbf{4}$

Answer: **4**

EXAMPLE 4. Find the value of $3y-7$ for $y=5$. When two operations are used, do any multiplication or division before addition or subtraction.

Step 1. Substitute $3y-7 = 3(5)-7$

Step 2. Solve $15-7 = \mathbf{8}$

Answer: **8**

Find the value of each algebraic expression below.

1. $x + 9$ for $x = -4$

2. $y - 12$ for $y = +3$

3. $\frac{x}{5}$ for $x = 35$

4. $6 - \frac{y}{3}$ for $y = -18$

5. $3ab + b$ for $a = +4$, $b = +2$

6. $y + 7$ for $y = 1$

7. $6 - z$ for $z = -7$

8. $6z$ for $z = -3$

9. $-\frac{n}{7} + 14$ for $n = 28$

10. $xy - \frac{y}{x}$ for $x = -6$, $y = -12$

11. $8 + z$ for $z = -5$

12. $x - 7$ for $x = 4$

13. xy for $x = -2$, $y = +4$

14. $4x - \frac{x}{3}$ for $x = 6$

15. $\frac{a}{b} - ab$ for $a = 14$, $b = -2$

Find the value of more complicated expressions by performing operations in the following order:

(1) Evaluate (solve) expressions within parentheses.
(2) Find powers.
(3) Do multiplication and division.
(4) Do addition and subtraction.

EXAMPLE 1. Evaluate $4(x - y)$ for $x = 6$ and $y = 3$. $4(x - y)$
Step 1. Substitute 6 for x and 3 for y. $= 4(6 - 3)$
Step 2. Evaluate the parentheses.
$(6 - 3) = (3)$
Step 3. Do multiplication. $= 4(3)$
Answer: **12** $=$ **12**

EXAMPLE 2. Evaluate $2(x - y)^2 - 4x$ for $x = 3$, $y = -2$ $2(x - y)^2 - 4x$
Step 1. Substitute 3 for x and -2 for y. $= 2(3 - (-2))^2 - 4(3)$
Step 2. Evaluate the parentheses.
$(3 - (-2)) = (3 + 2) = (5)$ $= 2(5)^2 - 4(3)$
Step 3. Evaluate the power.
$(5)^2 = 25$ $= 2(25) - 4(3)$
Step 4. Do multiplication.
$2(25) = 50$, and $4(3) = 12$
Step 5. Do subtraction. $= 50 - 12$
Answer: **38** $=$ **38**

Find the value of the following algebraic expressions.

16. $3(x + 7)$ for $x = 2$

17. $-4(y - 5)$ for $y = 3$

18. $-7(a - b)$ for $a = 2$, $b = 3$

19. $5(2x + y)$ for $x = 3$, $y = -4$

20. $2(z+3) - 5z$ for $z = -1$

21. $3(a - 2) + 3b$ for $a = +4$, $b = -3$

22. $4x^2$ for $x = 2$

23. $\frac{22}{7} z^2$ for $z = 14$

24. $(-a)^2 - 4$ for $a = 3$

25. $2x^3 - 5x$ for $x = -2$

26. $x^2 + y^2$ for $x = 4$, $y = -3$

27. $x^3 - z^2$ for $x = 3$, $z = 0$

In an expression involving division, the division bar separates the problem into two parts. Evaluate the numerator and the demoninator separately before dividing. Be sure to follow the order of operations shown on page 37.

EXAMPLE: Find the value of $\frac{3(y - 4)^2}{2(y - 1)}$ for $y = 6$.

$$\frac{3(y - 4)^2}{2(y - 1)}$$

Step 1. Substitute 6 for y. $= \frac{3(6 - 4)^2}{2(6 - 1)}$

Step 2. Evaluate both parentheses. $(6 - 4) = 2$ $(6 - 1) = 5$ $= \frac{3(2)^2}{2(5)}$

Step 3. Evaluate the power. $(2)^2 = 4$ $= \frac{3(4)}{2(5)}$

Step 4. Do multiplication. $3(4) = 12$ $2(5) = 10$ $= \frac{12}{10}$

Step 5. Simplify the improper fraction. $= \frac{6}{5} = 1\frac{1}{5}$

Answer: $1\frac{1}{5}$

Find the value of each algebraic expression below.

28. $\frac{2(x + 3)^2}{5(x - 1)}$ for $x = +2$

29. $\frac{3(m - n)^2}{2(m + n)}$ for $m = -1$, $n = +3$

30. $\frac{4z(z - 4)^2}{5z^2}$ for $z = -1$

31. $\frac{-2y(y - 1)^2}{3z(z + 1)^2}$ for $y = 2$, $z = 1$

EVALUATING FORMULAS

A *formula* is a mathematical rule. It is an algebraic expression that can be solved to find a particular quantity. For example, the formula that is used to find the area of a rectangle is: Area = length times width or $A = lw$.

In a formula, the quantity that you are trying to find is usually written to the left of the equal sign. The algebraic expression you need to solve is usually written to the right.

To find a value from a formula, substitute numbers for variables in the algebraic expression and do the arithmetic.

EXAMPLE: Using the formula $A = lw$, find the area of a rectangle when the length is 8 feet and the width is 5 feet.

Substitute 8 for l and 5 for w in the formula $A = lw$

$A = 8 \cdot 5 = 40$

Answer: $A =$ **40 square feet**

Note: In area problems, the answers will be in square feet, square inches, etc.
In volume problems, the answers will be in cubic feet, cubic inches, etc.

Evaluate each formula. (You don't need to know any geometry to solve these problems.) Write the answer on the line provided.

1. ***Description:*** Perimeter (distance around) of a rectangle
Formula: $P = 2(l + w)$
Variables: l = length
w = width
Find P when $l = 7$ feet
$w = 4$ feet

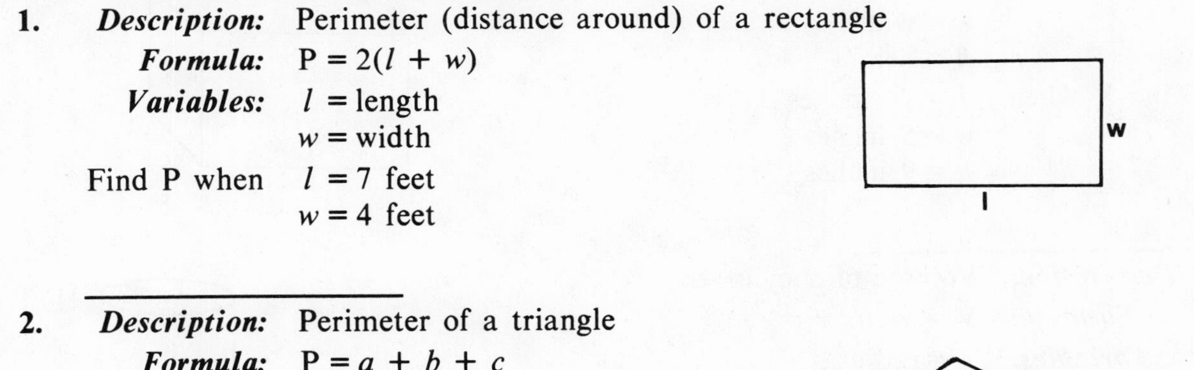

____

2. ***Description:*** Perimeter of a triangle
Formula: $P = a + b + c$
Variables: a, b, c = sides of triangle
Find P when $a = 6$ feet
$b = 9$ feet
$c = 13$ feet

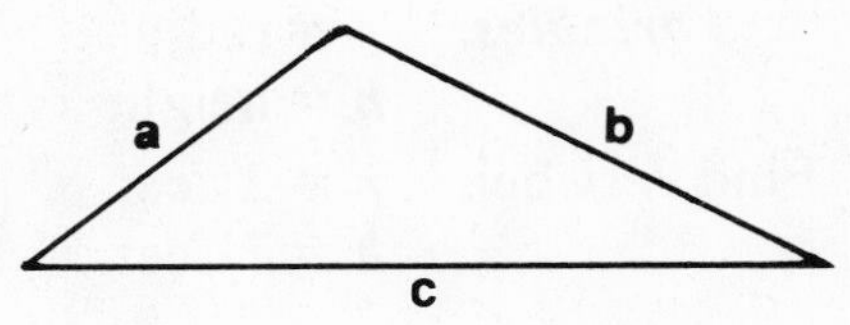

____

3. ***Description:*** Circumference (distance around) of a circle
Formula: $C = 2\pi r,\ \pi = \frac{22}{7}$
Variables: r = radius
Find C when $r = 7$ inches

____

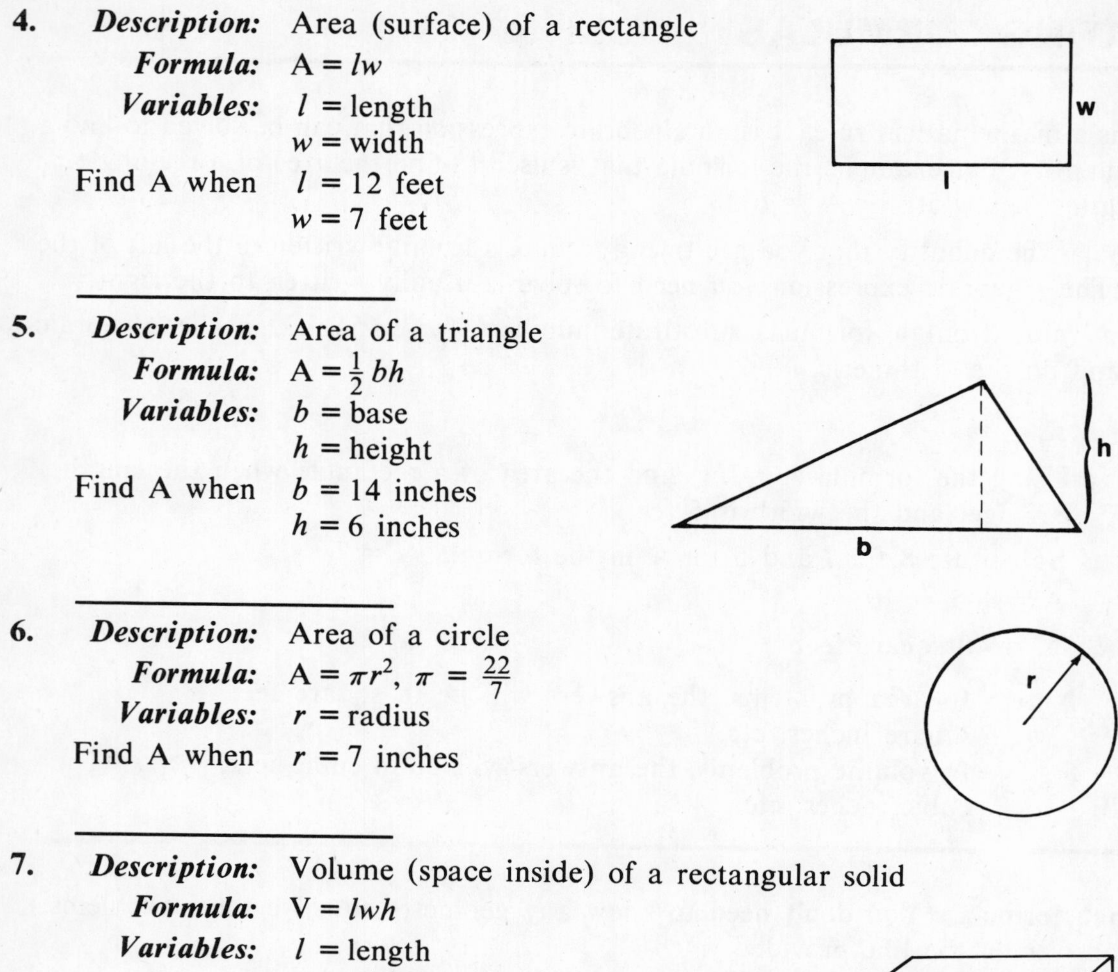

4. ***Description:*** Area (surface) of a rectangle

Formula: $A = lw$

Variables: l = length
w = width

Find A when l = 12 feet
w = 7 feet

____

5. ***Description:*** Area of a triangle

Formula: $A = \frac{1}{2} bh$

Variables: b = base
h = height

Find A when b = 14 inches
h = 6 inches

____

6. ***Description:*** Area of a circle

Formula: $A = \pi r^2$, $\pi = \frac{22}{7}$

Variables: r = radius

Find A when r = 7 inches

____

7. ***Description:*** Volume (space inside) of a rectangular solid

Formula: $V = lwh$

Variables: l = length
w = width
h = height

Find V when l = 14 inches
w = 5 inches
h = 9 inches

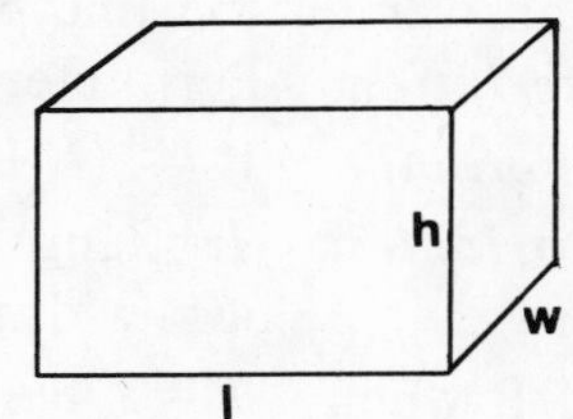

____

8. ***Description:*** Volume of a cylinder

Formula: $V = \pi r^2 h$, $\pi = \frac{22}{7}$

Variables: r = radius
h = height

Find V when r = 2 feet
h = 7 feet

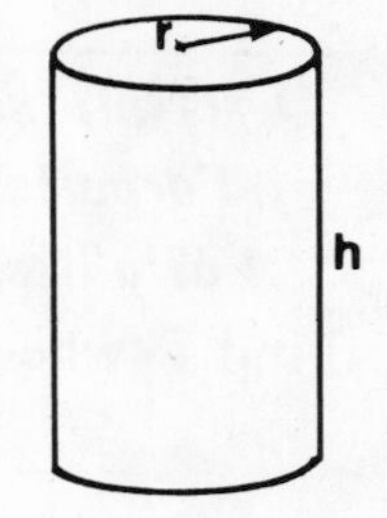

____

9. ***Description:*** Volume of a sphere

Formula: $V = \frac{4}{3}\pi r^3$, $\pi = \frac{22}{7}$

Variables: r = radius

Find V when r = 3 inches

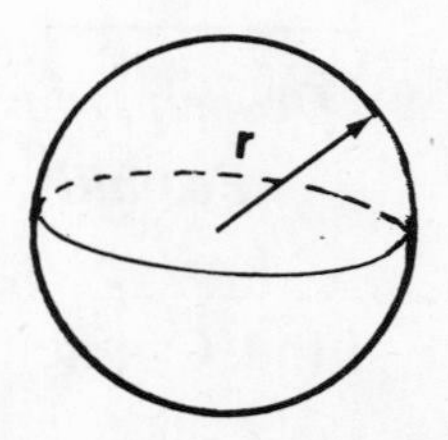

____

FORMULAS: APPLYING YOUR SKILLS

Using the formulas on the previous two pages, solve the problems below.

Use the perimeter formulas for finding the distance around (use circumference for a circle), use area formulas for finding the surface, and use volume formulas for finding the space inside an object.

1. George is enclosing a rectangular window with molding. How many feet of molding are needed if the window measures 5 feet long by 4 feet wide?

____

2. A school playground in the shape of a triangle has sides of 450 feet, 375 feet, and 574 feet. What is the distance around the playground?

____

3. How many feet of fence are needed to enclose a circular garden that has a radius of 28 feet?

____

4. Mary is putting new tile on her kitchen floor. How many square feet of tile does Mary need if the floor measures 15 feet long by 12 feet wide?

____

5. A triangular roof gable is 36 feet across and 10 feet high. Find the area of this gable.

____

6. Find the area of a circular garden that has a radius of 7 meters.

____

7. Carlos rented a moving van that has an enclosed section measuring 18 feet by 6 feet by 7 feet. How many cubic feet of storage space does the truck have?

____

ALGEBRAIC EXPRESSIONS SKILLS INVENTORY

Problems 1-5: Express each algebraic expression in words.

1. $12 + y$ ______________________ **2.** $x - 9$ ______________________

3. $15a$ ______________________ **4.** $\frac{13}{y}$ ______________________

5. $-9(2x + 4)$ ______________________

Problems 6-10: Write an algebraic expression for each word expression.

6. Seven subtracted from x ______________________

7. A number z divided by 9 ______________________

8. z plus nineteen ______________________

9. Eight m ______________________

10. Negative four times the quantity y minus 8 ______________________

Problems 11-18: Find the value of each algebraic expression.

11. $13 + 5x$ for $x = -3$

12. $7y - 8$ for $y = 6$

13. $\frac{-z}{3} + 9$ for $z = 15$

14. $\frac{x}{y} - 3xy$ for $x = 8$, $y = -2$

15. $-2\ (z + 7)$ for $z = -4$

16. $2a^2 - 4ab$ for $a = -2$, $b = +3$

17. $(m + n)(m - n)$ for $m = -4$, $n = 3$

18. $\frac{-2\ (x + y)^2}{(x - y)}$ for $x = -4$, $y = -1$

19. In the formula $V = \pi r^2 h$, find V when $r = 3$ feet and $h = 10$ feet. ($\pi = \frac{22}{7}$)

20. In the formula $I = prt$, find I when $p = \$550$, $r = 0.06$, and $t = 3$.

21. Using the area formula $A = lw$, find how many square feet of tile Howard needs to tile a kitchen that measures 15 feet by 12 feet.

22. Use the temperature formula $°C = \frac{5}{9}(°F - 32)$ to find the Celsius temperature when the Fahrenheit temperature is 104°.

ALGEBRAIC EXPRESSIONS INVENTORY CHART

Circle the number of any problem that you missed and be sure to review the appropriate practice page. A passing score is 19 correct answers.

Problem Number	Skill Area	Practice Page
1	writing algebraic expressions	34
2	writing algebraic expressions	34
3	writing algebraic expressions	34
4	writing algebraic expressions	34
5	writing algebraic expressions	35
6	writing algebraic expressions	34
7	writing algebraic expressions	34
8	writing algebraic expressions	34
9	writing algebraic expressions	34
10	writing algebraic expressions	35
11	evaluating algebraic expressions	36
12	evaluating algebraic expressions	36
13	evaluating algebraic expressions	36
14	evaluating algebraic expressions	36
15	evaluating algebraic expressions	37
16	evaluating algebraic expressions	37
17	evaluating algebraic expressions	37
18	evaluating algebraic expressions	38
19	evaluating formulas	39
20	evaluating formulas	39
21	applying formulas	41
22	applying formulas	41

If you missed more than 3 questions, you should review this chapter.

EQUATIONS

WHAT IS AN EQUATION?

An *equation* is a statement that two quantities are equal. You may see an equation written in words or in mathematical symbols:

Equation in words: Two plus three is equal to five
Equation in symbols: $2 + 3 = 5$

Note: The equal sign "=" is read "is equal to" or simply as "equals."

Equations are common in mathematics. In daily life, mathematical questions usually start as word problems and are solved as equations. We put the words into symbols because symbols are much easier to work with.

Each time you add, subtract, multiply, or divide numbers, you write an equation.

Familiar Equations
- Addition Equation: $32 + 19 = 51$
- Subtraction Equation: $76 - 30 = 46$
- Multiplication Equation: $43 \times 17 = 731$
- Division Equation: $\frac{80}{5} = 16$

In algebra, an equation may have a letter in place of a number. The letter stands for a number whose value is not yet known.

Algebraic Equations
- Addition Equation: $x + 23 = 47$
- Subtraction Equation: $y - 52 = 15$
- Multiplication Equation: $9a = 81$
- Division Equation: $\frac{b}{5} = 12$

Read an algebraic equation as follows:

$x + 23 = 47$ is read "*x* plus 23 is equal to 47"

$y - 52 = 15$ is read "*y* minus 52 equals 15"

$9a = 81$ is read "nine times *a* equals 81"

$\frac{b}{5} = 12$ is read "*b* divided by 5 is equal to 12"

Express each equation below in words. More than one expression is possible for each equation.

1. $x + 52 = 72$ ______________________

2. $y - 21 = 13$ ______________________

3. $7a = 147$ ______________________

4. $\frac{z}{6} = 16$ ______________________

UNKNOWNS AND CHECKING UNKNOWNS

As you've seen, an algebraic equation contains a letter in place of a number. We call this letter the *unknown* because it represents a number that we don't know. An unknown is sometimes called a variable.

Solving an algebraic equation means finding the value of the unknown that makes the equation a true statement. The *solution* is the value of the unknown that solves the equation.

On the pages ahead, you'll learn to solve each type of algebraic equation. On this page, you'll learn to check a possible value for the unknown to make sure that you've solved an equation correctly.

To check if a possible value for the unknown is the solution of an equation, follow these two steps:

Step 1. Substitute the value for the unknown into the original equation.

Step 2. Simplify (do the arithmetic) and compare each side of the equation.

EXAMPLE 1. Is $y = 5$ the solution for $3y - 9 = 6$?

Step 1. Substitute 5 for y. $3(5) - 9 = 6$?

Step 2. Simplify and compare. $15 - 9 = 6$?

$6 = 6$?

Since **6 = 6, y = 5 is a solution** of the equation.

EXAMPLE 2. Is $x = 23$ the solution for $x - 7 = 14$?

Step 1. Substitute 23 for x. $(23) - 7 = 14$?

Step 2. Simplify and compare. $16 = 14$?

Since 16 is not equal to 14, **x = 23 is not a solution** of the equation.

Check if the possible value for the unknown is the solution to each equation. Circle "Yes" if the value is the solution, and circle "No" if it is not.

1. $x + 3 = 9$ Try $x = 4$ Yes No

2. $y - 12 = 37$ Try $y = 49$ Yes No

3. $3z = 39$ Try $z = 13$ Yes No

4. $\frac{a}{15} = 3$ Try $a = 54$ Yes No

5. $\frac{3}{4}m = 3$ Try $m = 4$ Yes No

SOLVING AN ALGEBRAIC EQUATION

An algebraic equation is very much like a balance. The left side of the equation "balances" or "equals" the right side of the equation.

The quantities of the addition equation $x + 3 = 5$ can be represented as weights on the sides of the balance. On the left side are the unknown weight x and 3 unit weights. On the right side are 5 unit weights.

To find the unknown weight x, *remove an equal number of weights from each side* until x stands alone on the left side. Removing an equal number from each side insures that the equation still balances after the weights are removed.

You have solved the equation when x stands alone on the left side and the value that balances x stands alone on the right side.

As shown to the right, the equation $x + 3 = 5$ is solved by *subtracting 3 units from each side of the balance.* That leaves two units to balance the quantity x. Therefore, the solution is $x = 2$.

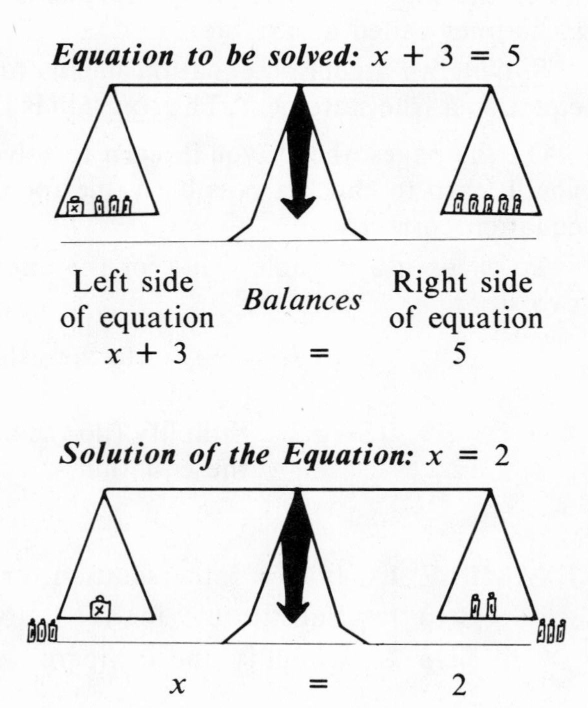

You can see that subtraction was used to solve the addition equation $x + 3 = 5$. Subtracting $+3$ from each side removes the $+3$ from the left side and leaves x standing alone. Because subtraction "undoes" addition, subtraction is called the *inverse* (opposite) of addition.

As you'll see in the pages ahead, all algebraic equations are solved by using *inverse operations*. An inverse operation "undoes" an operation and leaves the unknown standing alone.

Operation		**Inverse Operation**	**Solving An Equation**
addition	⟷	subtraction	Use subtraction to solve an addition equation.
subtraction	⟷	addition	Use addition to solve a subtraction equation.
multiplication	⟷	division	Use division to solve a multiplication equation.
division	⟷	multiplication	Use multiplication to solve a division equation.

The example $x + 3 = 5$ also points out the most important rule to remember as you use inverse operations to solve equations:

RULE: Any inverse operation performed on one side of the equation must also be performed on the other side of the equation.

SOLVING AN ADDITION EQUATION

To solve an algebraic equation, you must get the unknown alone on one side of the equation. The number that the unknown equals is the solution to the equation.

To solve an *addition equation,* one in which a number is being added to an unknown, subtract that number from each side of the equation. Since a number subtracted from itself is zero, this results in the unknown being left by itself on one side of the equation.

Remember, subtraction is the inverse of addition.

EXAMPLE: Solve for x in $x + 4 = 11$

Step 1. Subtract 4 from each side.

Solve: $x + 4 = 11$

$\underbrace{x + 4 - 4}_{0} = 11 - 4$

Step 2. Simplify both sides of the equation. Since $4 - 4$ equals 0, the left side of the equation leaves x standing alone.

$x = 7$

Answer: $x = 7$

Check by substituting 7 for x.

Check: $7 + 4 = 11$

$11 = 11$

Solve each addition equation. Check your answers in the space provided.

Examples

Solve: $x + 2 = 9$

$\underbrace{x + 2 - 2}_{0} = 9 - 2$

$x = 7$

Check: $7 + 2 = 9$

$9 = 9$

Solve: $y + 6 = 17$

$\underbrace{y + 6 - 6}_{0} = 17 - 6$

$y = 11$

Check: $11 + 6 = 17$

$17 = 17$

Solve: $6 + z = 4$

$\underbrace{6 - 6}_{0} + z = 4 - 6$

$z = -2$

Check: $6 + (-2) = 4$

$4 = 4$

1. $y + 3 = 7$

2. $a + 6 = 8$

3. $z + 5 = 19$

4. $b + 3 = 14$

5. $7 + a = 3$

6. $12 + x = 5$

SOLVING A SUBTRACTION EQUATION

To solve a *subtraction equation*, one in which a number is being subtracted from an unknown, add that number to each side of the equation. Since a number added to its opposite is zero, this results in the unknown being left alone on one side of the equation.

Remember, addition is the inverse of subtraction.

EXAMPLE: Solve for y in $y - 6 = 3$

Step 1. Add 6 to each side.

Solve: $y - 6 = 3$

$y \underbrace{- 6 + 6}_{0} = 3 + 6$

Step 2. Simplify both sides of the equation. (Since $-6 + 6$ equals 0, the left side of the equation leaves y standing alone.)

$y + 0 = 9$

$y = \mathbf{9}$

Answer: $y = \mathbf{9}$

Check by substituting 9 for y.

Check: $9 - 6 = 3$

$3 = 3$

Solve each subtraction equation. Check each answer in the space provided.

Examples

Solve: $x - 7 = 4$

$x \underbrace{- 7 + 7}_{0} = 4 + 7$

$x = 11$

Check: $11 - 7 = 4$

$4 = 4$

Solve: $y - 9 = 13$

$y \underbrace{- 9 + 9}_{0} = 13 + 9$

$y = 22$

Check: $22 - 9 = 13$

Solve: $z - 4 = -2$

$z \underbrace{-4 + 4}_{0} = -2 + 4$

$z = 2$

Check: $2 - 4 = -2$

$-2 = -2$

1. $y - 4 = 8$

2. $z - 2 = 9$

3. $a - 21 = 34$

4. $b - 16 = 49$

5. $x - 12 = -6$

6. $c - 14 = -23$

SOLVING A MULTIPLICATION EQUATION

To solve a *multiplication equation*, one in which the unknown is multiplied by a number (called a *coefficient*), divide each side of the equation by that number. A number divided by itself results in a coefficient of 1 and leaves the unknown alone on one side of the equation.

Remember, division is the inverse of multiplication.

EXAMPLE: Solve for a in $5a = 20$

Step 1. Divide each side by 5.
Simplify each side of the equation.

Step 2. Since $\frac{5}{5}$ is 1, write a on one side of the equation.
(The coefficient 1 is not written in algebra.)
Solve for a by dividing 20 by 5.

Answer: $\mathbf{a = 4}$

Check by substituting 4 for a.

Solve: $5a = 20$

$\frac{{}^1\cancel{5}a}{\cancel{5}_1} = \frac{20}{5}$

$\mathbf{a = 4}$

Check: $5(4) = 20$

Solve each multiplication equation. Check each answer as indicated.

Examples

Solve: $6x = 24$

$\frac{{}^1\cancel{6}x}{\cancel{6}_1} = \frac{24}{6}$

$\mathbf{x = 4}$

Check: $6 \cdot 4 = 24$

$24 = 24$

Solve: $-4x = 12$

$\frac{{}^1\cancel{-4}x}{\cancel{-4}_1} = \frac{12}{-4}$

$\mathbf{x = -3}$

Check: $-4(-3) = 12$

$12 = 12$

Solve: $-y = 13$

$-1 \cdot y = 13$

$\frac{{}^1\cancel{-1} \cdot y}{\cancel{-1}_1} = \frac{13}{-1}$

$\mathbf{y = -13}$

Check: $-(-13) = 13$

$13 = 13$

1. $5y = 35$

2. $8a = 2$

3. $-7y = 28$

4. $-15b = 10$

5. $-z = 9$

6. $-x = -17$

SOLVING A DIVISION EQUATION

To solve a *division equation*, one in which the unknown is divided by a number, multiply each side of the equation by that number. The numbers will "cancel out," leaving a coefficient of 1. Since we do not write a coefficient of 1 in algebra, the unknown is left alone on one side of the equation.

Remember, multiplication is the inverse of division.

EXAMPLE: Solve for z in $\frac{z}{3} = 4$

Step 1. Multiply each side by 3.

Step 2. Since $\frac{3}{3}$ is 1, write z on one side of the equation. (The coefficient 1 is not written.) Solve for z by multiplying 4 times 3.

Answer: $\mathbf{z = 12}$

Check by substituting 12 for z.

Solve: $\frac{z}{3} = 4$

$\frac{{}^{1}\cancel{3}z}{\cancel{3}_{1}} = 4(3)$

$\mathbf{z = 12}$

Check: $\frac{12}{3} = 4$

$4 = 4$

Solve each division equation. Check your answers in the space provided.

Examples

Solve: $\frac{x}{6} = 2$

$\frac{{}^{1}\cancel{6}x}{\cancel{6}_{1}} = 2(6)$

$\mathbf{x = 12}$

Check: $\frac{12}{6} = 2$

$2 = 2$

Solve: $\frac{a}{8} = \frac{1}{2}$

$\frac{{}^{1}\cancel{8}a}{\cancel{8}_{1}} = \frac{1}{2}(8)$

$\mathbf{a = 4}$

Check: $\frac{4}{8} = \frac{1}{2}$

$\frac{1}{2} = \frac{1}{2}$

Solve: $\frac{y}{3} = -5$

$\frac{{}^{1}\cancel{3}y}{\cancel{3}_{1}} = (-5)(3)$

$\mathbf{y = -15}$

1. $\frac{y}{5} = 7$

2. $\frac{m}{9} = 8$

3. $\frac{b}{6} = \frac{2}{3}$

4. $\frac{c}{4} = \frac{1}{2}$

5. $\frac{x}{7} = -8$

6. $\frac{z}{9} = -7$

SOLVING A FRACTIONAL EQUATION

An equation with a fraction multiplying the unknown is called a *fractional equation.*

For example, $\frac{3x}{4} = 9$ is a fractional equation. The preferred way to write the fraction is with the fraction bar under the unknown. However, you may also see it written with the fraction in front. For example, $\frac{3x}{4}$ is the preferred way to write $\frac{3}{4}x$.

A negative fraction can be written with the negative sign either in the numerator or denominator. However, the preferred way is to write the negative sign in front of the fraction. For example, $-\frac{5x}{6}$ is the preferred way to write either $\frac{-5x}{6}$ or $\frac{5x}{-6}$.

To solve a fractional equation, multiply each side of the equation by the *reciprocal* of the fraction. The reciprocal is found by inverting (turning upside-down) the fraction. For example, the reciprocal of $\frac{3}{4}$ is $\frac{4}{3}$. Multiplying by the reciprocal "cancels out" the coefficient and leaves the unknown standing alone on one side of the equation.

EXAMPLE: Solve for x in $\frac{3x}{4} = 9$.

Solve: $\frac{3x}{4} = 9$

Step 1. Multiply each side by $\frac{4}{3}$.

$\left(\frac{\cancel{4}^{1}}{\cancel{3}_{1}}\right)\frac{\cancel{3}^{1}x}{\cancel{4}_{1}} = {}^{3}\cancel{9}\left(\frac{4}{\cancel{3}_{1}}\right)$

Step 2. Since $\left(\frac{\cancel{4}^{1}}{\cancel{3}_{1}}\right)\left(\frac{\cancel{3}^{1}}{\cancel{4}_{1}}\right) = 1$, write x on the left side of the equation. Solve for x by multiplying $9\left(\frac{4}{3}\right)$.

$\mathbf{x = 12}$

Answer: $\mathbf{x = 12}$

Check by substituting 12 for x.

Check: $\frac{3 \cdot 12}{4} = 9$

$9 = 9$

Solve each fractional equation. Check each answer as indicated.

Examples

Solve: $\frac{2y}{3} = 8$

$\left(\frac{\cancel{3}^{1}}{\cancel{2}_{1}}\right)\frac{\cancel{2}^{1}y}{\cancel{3}_{1}} = {}^{4}\cancel{8}\left(\frac{3}{\cancel{2}_{1}}\right)$

$y = 12$

Check: $\frac{2 \cdot 12}{3} = 8$

$8 = 8$

Solve: $-\frac{4z}{5} = 12$

$\left(-\frac{\cancel{5}^{1}}{\cancel{4}_{1}}\right)\left(-\frac{\cancel{4}^{1}z}{\cancel{5}_{1}}\right) = {}^{3}\cancel{12}\left(-\frac{5}{\cancel{4}_{1}}\right)$

$z = -15$

Check: $\frac{-4(-15)}{5} = 12$

$12 = 12$

1. $\frac{5z}{6} = 10$

2. $\frac{3a}{7} = 21$

3. $-\frac{5x}{8} = 15$

4. $-\frac{4y}{9} = -8$

SOLVING AN EQUATION WITH ONE OPERATION

You've now learned how to solve basic algebraic equations. Each of these equations involves a single operation, and each is solved by performing the inverse operation on the other side of the equation.

Increase your algebra skills by reviewing the basic equations on this page.

Addition Equations: *Solve by subtraction.*

1. $x + 8 = 10$ **2.** $5 + y = 9$ **3.** $a + 5 = -4$ **4.** $z + 5 = 8$

Subtraction Equations: *Solve by addition.*

5. $y - 7 = 12$ **6.** $x - 12 = 13$ **7.** $b - 3 = -7$ **8.** $c - 9 = 6$

Multiplication Equations: *Solve by division.*

9. $9z = 72$ **10.** $4x = -32$ **11.** $13y = 143$ **12.** $-a = 9$

Division Equations: *Solve by multiplication.*

13. $\frac{x}{3} = 9$ **14.** $\frac{y}{7} = 8$ **15.** $\frac{z}{4} = 7$ **16.** $\frac{a}{6} = -4$

Fractional Equations: *Solve by multiplying by the reciprocal of the fractional coefficient.*

17. $\frac{2x}{5} = 6$ **18.** $\frac{-3z}{7} = 9$ **19.** $\frac{4y}{9} = 16$ **20.** $\frac{7a}{4} = -14$

Solve each equation for the unknown.

21. $y - 5 = 7$ **22.** $\frac{z}{8} = -3$ **23.** $9a = 81$ **24.** $6 + y = 13$

25. $\frac{3x}{5} = 12$ **26.** $a + 12 = 9$ **27.** $z - 14 = -3$ **28.** $\frac{x}{8} = 4$

29. $\frac{y}{12} = 5$ **30.** $-4x = 16$ **31.** $-z = -9$ **32.** $\frac{3}{4}a = -18$

33. $x - (-4) = 5$ **34.** $y + 8 = 6$ **35.** $\frac{b}{-3} = -5$ **36.** $-7c = -35$

37. $y + (-6) = 3$ **38.** $\frac{x}{8} = -\frac{3}{4}$ **39.** $-\frac{2y}{3} = 6$ **40.** $-x = 12$

41. $5e = -19$ **42.** $y + (-4) = -6$ **43.** $z - (-3) = -6$

44. $\frac{y}{-5} = -9$

SOLVING AN EQUATION WITH ONE OPERATION: APPLYING YOUR SKILLS

Equations may be used to solve word problems. To solve a word problem, read the whole problem carefully and then follow these three steps:

Step 1. Represent the unknown with a letter.
Step 2. Write an equation that represents the problem.
Step 3. Solve the equation for the unknown.

EXAMPLE 1. Seven times a number is equal to 147. What is the number?

Step 1. Let x equal the unknown number.
Step 2. Write an equation for the problem: $7x = 147$
Step 3. Solve the equation.
Divide each side by 7.

Solve:

$$7x = 147$$

$$\frac{{}^{1}\cancel{7}x}{\cancel{7}_{1}} = \frac{{}^{21}\cancel{147}}{\cancel{7}_{1}}$$

$$\mathbf{x = 21}$$

Answer: $\mathbf{x = 21}$

The unknown number is 21.

Check:

$$7(21) = 147$$

$$147 = 147$$

Example 2 shows how to set up a word or story problem in algebra. Study this carefully.

EXAMPLE 2. Bill saves $\frac{1}{8}$ of his monthly pay check. If his monthly savings is \$92, how much does he earn each month?

Step 1. Let x = monthly income because this is the unknown quantity that you must find.
\$92 = monthly savings
$\frac{1}{8}$ = fraction saved

Step 2. Write an equation for the problem.
Fraction saved times income = savings
$\frac{1}{8}x = \$92$

Step 3. Solve the equation.
Multiply each side by 8.

Solve:

$$\frac{1}{8}x = \$92$$

$${}^{1}(\cancel{8})\frac{1}{\cancel{8}_{1}}x = \$92(8)$$

$$\mathbf{x = \$736}$$

Answer: $\mathbf{x = \$736}$

Bill earns \$736 each month.

Check:

$$\frac{1}{8}(736) = 92$$

$$92 = 92$$

Solve the following problems.

1. If 8 is added to a certain number, the sum is 19. What is the number?

____

2. When a certain number is decreased by 12, the result is 9. Find the number.

____

3. Four times an unknown number is equal to 28. What is the unknown number?

____

4. If a large number is divided by 12, the answer is 121. What is the large number?

____

5. Joan pays $\frac{1}{12}$ of her total monthly income in property taxes. If she paid \$45 last month in taxes, what was her monthly income?

____

6. Jim pays $\frac{2}{7}$ of his monthly earnings for the rent of his truck. If his truck rental averages \$220 a month, what is his average monthly income?

____

7. During a sale, Allen paid 80% (0.80) of the original price of a coat. If Allen paid \$76, what was the original price?

____

8. As a salesperson, Diane earns an 8% (0.08) commission on each sale. During December, Diane earned \$758 in commissions. What was her total for monthly sales? (Commission is a percent of sales.)

____

SOLVING AN EQUATION WITH SEVERAL OPERATIONS

A single equation may contain several operations. More than one of the operations of addition, subtraction, multiplication, or division may be needed to find a solution.

To solve an equation with more than one operation, follow these two rules:

RULE 1: Do addition or subtraction first.

a. If a number is added to the unknown, subtract that number from each side of the equation.

b. If a number is subtracted from the unknown, add that number to each side of the equation.

RULE 2: Do multiplication or division last.

a. If the unknown is multiplied by a number, divide each side of the equation by that number.

b. If the unknown is divided by a number, multiply each side of the equation by that number.

EXAMPLE 1. Solve for x in $2x - 7 = 19$

Step 1. Add 7 to each side.
Remember that $-7+7 = 0$. We do not need to write the 0 in the equation.

Step 2. Divide each side by 2.
Remember that $\frac{2}{2} = 1$. We do not need to write 1 as a coefficient of x.

Answer: $\mathbf{x = 13}$

Check by substituting 13 for x.

Solve:

$$2x - 7 = 19$$
$$2x - 7 + 7 = 19 + 7$$
$$2x = 26$$
$$\frac{{}^1\cancel{2}x}{\cancel{2}_1} = \frac{26}{2}$$
$$\mathbf{x = 13}$$

Check:

$$2(13) - 7 = 19$$
$$26 - 7 = 19$$
$$19 = 19$$

EXAMPLE 2. Solve for z in $\frac{z}{4} + 6 = 22$

Step 1. Subtract 6 from each side.

Step 2. Multiply each side by 4

Answer: $\mathbf{z = 64}$

Check by substituting 64 for z.

Solve:

$$\frac{z}{4} + 6 = 22$$
$$\frac{z}{4} + 6 - 6 = 22 - 6$$
$$\frac{z}{4} = 16$$
$$\frac{{}^1\cancel{4}z}{\cancel{4}_1} = 16(4)$$
$$\mathbf{z = 64}$$

Check:

$$\frac{64}{4} + 6 = 22$$
$$16 + 6 = 22$$
$$22 = 22$$

Solve each equation below. Check each answer in the space provided.

Examples

Solve: $5y + 4 = 19$

$5y + 4 - 4 = 19 - 4$

$5y = 15$

$\frac{\cancel{5}y}{\cancel{5}} = \frac{15}{5}$

$y = 3$

Check: $5(3) + 4 = 19$

$15 + 4 = 19$

$19 = 19$

Solve: $3x - 9 = 3$

$3x - 9 + 9 = 3 + 9$

$3x = 12$

$\frac{\cancel{3}x}{\cancel{3}} = \frac{12}{3}$

$x = 4$

Check: $3(4) - 9 = 3$

$12 - 9 = 3$

$3 = 3$

Solve: $\frac{y}{4} + 5 = 3$

$\frac{y}{4} + 5 - 5 = 3 - 5$

$\frac{y}{4} = -2$

$\frac{\cancel{4}y}{\cancel{4}} = (-2)(4)$

$y = -8$

Check: $\frac{-8}{4} + 5 = 3$

$-2 + 5 = 3$

$3 = 3$

Solve: $7 - 3b = -11$

$+7 - 7 - 3b = -11 - 7$

$-3b = -18$

$\frac{-\cancel{3}b}{-\cancel{3}} = \frac{-18}{-3}$

$b = 6$

1. $3x + 11 = 20$

2. $2z + 9 = 1$

3. $4z - 8 = 24$

4. $-3a - 4 = 17$

5. $\frac{z}{2} + 6 = 7$

6. $-\frac{x}{5} + 3 = 9$

7. $2 - 4z = -14$

8. $1 - 5y = 11$

LEARNING A SHORTCUT TO SOLVING EQUATIONS

To solve an equation, you must get the unknown standing alone on one side. In the two examples below, we'll show you a *shortcut* that saves a lot of writing. In the shortcut, you do not write numbers that add to 0, and you do not write coefficients that equal 1. Compare the shortcut with the long way we have been using.

Shortcut

EXAMPLE 1. Solve $3x + 9 = 30$

Step 1. Move the 9 to the other side of the equation and change its sign from addition to subtraction.

$3x = 30 - 9$
$3x = 21$

Step 2. Divide 21 by 3.

$x = \frac{21}{3}$

Answer: $\mathbf{x = 7}$

Long Way

Solve $3x + 9 = 30$

Step 1. Subtract 9 from each side.

$3x + 9 - 9 = 30 - 9$
$3x = 21$

Step 2. Divide each side by 3.

$\frac{\overset{1}{\cancel{3}}x}{\cancel{3}_1} = \frac{\overset{7}{\cancel{21}}}{\cancel{3}_1}$

Answer: $\mathbf{x = 7}$

Shortcut

EXAMPLE 2. Solve $\frac{y}{2} - 12 = 31$

Step 1. Move the 12 to the other side of the equation and change its sign from subtraction to addition.

$\frac{y}{2} = 31 + 12$
$\frac{y}{2} = 43$

Step 2. Multiply 43 by 2.

$y = 43 \cdot 2$

Answer: $\mathbf{y = 86}$

Long Way

Solve $\frac{y}{2} - 12 = 31$

Step 1. Add 12 to each side.

$\frac{y}{2} - 12 + 12 = 31 + 12$
$\frac{y}{2} = 43$

Step 2. Multiply each side by 2.

$\frac{\overset{1}{\cancel{2}}y}{\cancel{2}_1} = 43 \cdot 2$

Answer: $\mathbf{y = 86}$

The steps of the shortcut can be summed up as follows:

Step 1. If a number is added to (or subtracted from) an unknown, move the number to the other side of the equation and change its sign.

Step 2. If a number multiplies an unknown, solve for the unknown by dividing the other side of the equation by the number.

Step 3. If a number divides an unknown, solve for the unknown by multiplying the other side of the equation by the number.

We will use the shortcut in the rest of our work with equations in this chapter.

Solve the equations below using the shortcut. Check each answer on scratch paper.

Examples

$3x + 11 = 29$
$3x = 29 - 11$
$3x = 18$
$x = \frac{18}{3}$
$x = 6$

$-12 + 2y = 14$
$2y = 14 + 12$
$2y = 26$
$y = \frac{26}{2}$
$y = 13$

$\frac{2z}{3} + 4 = 10$
$\frac{2z}{3} = 10 - 4$
$\frac{2z}{3} = 6$
$z = \overset{3}{\cancel{6}}\left(\frac{3}{\cancel{2}_1}\right)$
$z = 9$

1. $4y + 9 = 37$

2. $7z - 6 = 29$

3. $-5 + 7a = 30$

4. $13 - 9x = -41$

5. $\frac{5x}{6} + 9 = 24$

6. $\frac{2a}{7} - 8 = 12$

Here are more equations to solve using the shortcut.

7. $2x + 5 = 7$

8. $2 + 3a = 11$

9. $4y + 1 = 13$

10. $3z - 7 = 8$

11. $-14 + 2x = 4$

12. $8x - 12 = -4$

13. $2 + 3b = -1$

14. $\frac{x}{2} + 7 = 9$

15. $\frac{z}{3} - 2 = 5$

SOLVING AN EQUATION WITH SEPARATED UNKNOWNS

An equation is made up of *terms*. Each term is a number standing alone or an unknown multiplied by a coefficient (a number and the sign that precedes it).

When an unknown appears in more than one term in an equation, the separate terms can be combined. Combine the coefficients according to the rules for combining signed numbers and attach the unknown.

For example, the terms below are combined as follows:

$5x + 2x = 7x$ since $5 + 2 = 7$
$7y - 4y = 3y$ since $7 - 4 = 3$
$8z + z = 9z$ since $8 + 1 = 9$
$6z - z = 5z$ since $6 - 1 = 5$

Note: When no number multiplies (is in front of) an unknown, the coefficient is understood to be one. Thus, the coefficient of z is $+1$, and the coefficient of $-z$ is -1.

To solve an equation in which the unknown appears in more than one term, combine the separate terms as your first step.

EXAMPLE: Solve for x in $4x - x + 2 = 17$

Step 1. Combine the x's.

Step 2. Subtract 2 from 17.

Step 3. Divide 15 by 3.

Answer: $\mathbf{x = 5}$

Check by substituting 5 for x.

Solve:
$$4x - x + 2 = 17$$
$$3x + 2 = 17$$
$$3x = 17 - 2$$
$$3x = 15$$
$$x = \frac{15}{3}$$
$$\mathbf{x = 5}$$

Check:
$$4(5) - 5 + 2 = 17$$
$$20 - 5 + 2 = 17$$
$$17 = 17$$

Solve the following equations.

Example

$2x + 3x = 25$
$5x = 25$
$x = \frac{25}{5}$
$\mathbf{x = 5}$

1. $3y + 5y = 32$

2. $5z + z = 24$

Examples

$$3a - a = 18 + 6$$
$$2a = 24$$
$$a = \frac{24}{2}$$
$$a = 12$$

$$2x + 4x - 9 = 15$$
$$6x - 9 = 15$$
$$6x = 15 + 9$$
$$6x = 24$$
$$x = \frac{24}{6}$$
$$x = 4$$

$$\frac{3y}{4} - \frac{y}{2} - 5 = 7$$
$$\frac{y}{4} - 5 = 7$$
$$\frac{y}{4} = 7 + 5$$
$$\frac{y}{4} = 12$$
$$y = 12(4)$$
$$y = 48$$

3. $7y - 3y = 23 + 9$

4. $13x - 8x = 17 + 8$

5. $9z - z - 5 = 11$

6. $12y - 11y - 9 = 17$

7. $\frac{3}{5}a - \frac{1}{5}a + 3 = 17$

8. $\frac{2}{3}x + \frac{1}{6}x - 4 = 11$

Solve each equation below. Remember to combine separated unknowns as your first step in solving the equation.

9. $5z + 7z = 48$

10. $3y - 5y = 15$

11. $4a + 3a - 4 = 10$

12. $4a - 2a = 7 + 13$

13. $\frac{1}{2}x + \frac{1}{4}x = 13 + 11$

14. $2y - y + 2 = 7$

15. $5b - 2b = 14 - 5$

16. $3z + 6z + 7 = 29 - 4$

17. $y - \frac{1}{2}y = 3$

TERMS ON BOTH SIDES OF AN EQUATION

Terms that contain an unknown can appear on both sides of an equation. Usually, we move all terms containing an unknown to the left side of the equation. To move a term containing an unknown from the right side of the equation to the left side, change its sign and place it next to the unknown on the left side.

RULE 1: If the term is preceded by a positive sign, remove the term from the right side and subtract it from the left side.

RULE 2: If the term is preceded by a negative sign, remove the term from the right side and add it to the left side.

EXAMPLE 1. Solve for x in $3x = 2x + 7$

Step 1. Subtract $2x$ from $3x$.

Step 2. Combine the x's.

Answer: $x = 7$

Check by substituting 7 for x.

Solve:

$$3x = 2x + 7$$
$$3x - 2x = 7$$
$$x = 7$$

Check:

$$3(7) = 2(7) + 7$$
$$21 = 14 + 7$$
$$21 = 21$$

EXAMPLE 2. Solve for y in $2y - 6 = -3y + 24$

Step 1. Add $3y$ to $2y - 6$.

Step 2. Combine the y's.

Step 3. Add 6 to 24.

Step 4. Divide 30 by 5.

Answer: $y = 6$

Check by substituting 6 for y.

Solve:

$$2y - 6 = -3y + 24$$
$$2y + 3y - 6 = 24$$
$$5y - 6 = 24$$
$$5y = 24 + 6$$
$$5y = 30$$
$$y = \frac{30}{5}$$
$$y = 6$$

Check:

$$2(6) - 6 = (-3)(6) + 24$$
$$12 - 6 = -18 + 24$$
$$6 = 6$$

Solve each equation below.

Example

$$4x = 3x + 6$$
$$4x - 3x = 6$$
$$x = 6$$

1. $5y = 2y + 9$

2. $7z = 5z + 18$

Examples

$$3z = 12 - z$$
$$3z + z = 12$$
$$4z = 12$$
$$z = \frac{12}{4}$$
$$z = 3$$

$$5x - 4 = 3x + 12$$
$$5x - 3x - 4 = 12$$
$$2x - 4 = 12$$
$$2x = 12 + 4$$
$$2x = 16$$
$$x = \frac{16}{2}$$
$$x = 8$$

3. $5a = -3a + 24$

4. $7x = -27 - 2x$

5. $9b - 7 = 5b + 25$

6. $8y - 9 = -y + 9$

Solve each equation below. Check any problem you're unsure of on scratch paper.

7. $12x = 7x + 20$

8. $9y = - 2y + 33$

9. $3z = z - 18$

10. $4a + 17 = a - 13$

11. $6b - 12 = 2b - 4$

12. $z = \frac{1}{2}z + 4$

13. $5y + 9 = -2y + 30$

14. $14 + 2x = x - 9$

15. $11z - 3 = 9z$

SOLVING AN EQUATION: APPLYING YOUR SKILLS

Study the following examples carefully before trying the word problems. Notice how the equation is written to represent the information contained in the problem.

EXAMPLE 1. Six times a number plus 7 is equal to 55. What is the number?

Step 1. Let x equal the unknown number.

Step 2. Write an equation for the problem.
$6x + 7 = 55$

Step 3. Solve the equation
a) Subtract 7 from 55.
b) Divide 48 by 6.

Answer: **The number is 8.**

Solve:
$$6x + 7 = 55$$
$$6x = 55 - 7$$
$$6x = 48$$
$$x = \frac{48}{6}$$
$$\mathbf{x = 8}$$

EXAMPLE 2. Jack and Steve do yard work. Because Jack provides the truck, gas, and yard equipment, he receives twice the money that Steve does. If they collect $540, how much does each receive?

Step 1. Let x = Steve's share
$2x$ = Jack's share (We know that Jack receives twice Steve's share.)

Step 2. Write an equation for the problem.
Jack's share plus Steve's share = $540
$2x + x = 540$

Step 3. Solve the equation.
a) Combine the x's.
b) Divide 540 by 3.

Answer: **x = $180 Steve's share**
$2x$ = $360 Jack's share

Solve:
$$2x + x = 540$$
$$3x = 540$$
$$x = \frac{540}{3}$$
$$\mathbf{x = \$180}$$
$$\mathbf{2x = 2(180) = \$360}$$

Solve the problems below.

1. Eight times a number plus 9 is equal to 73. What is the number?

____

2. Five times a number minus 7 is equal to three times the same number plus 19. What is the number?

____

3. An unknown number divided by 3 is equal to the same number divided by 6, plus 4. What is the unknown number?

____

4. Two-thirds of a number plus one-sixth of the same number is equal to 25. What is the number?

____

5. Susan and Terry run a day-care center. Since they use Susan's house, it was agreed that her share is to be twice Terry's share. If they earn $225, how much is each person's share?

____

6. Jorge and Manuel had a roofing business. Jorge, as owner of the materials, received 3 dollars for every 2 dollars Manuel received. On a job that paid $750, what amount did each receive?

____

7. Lucy earns $400 a month in salary and she receives a commission of $18 for each appliance she sells. If last month Lucy earned a total of $886, how many appliances did she sell?

____

8. Joe has money in a savings account. If he adds $50 a month for 6 months, he will have three times the amount he has now, not counting the interest. How much is in Joe's account now?
(*Hint:* Let x = amount he has now; $x + 300$ = amount he will have in 6 months)

____

SOLVING AN EQUATION WITH PARENTHESES

Parentheses are commonly used in algebraic equations. Parentheses are used to identify terms that are to be multiplied by another term, usually a number.

The first step in solving an equation is to remove the parentheses by multiplication. Then, combine separated unknowns and solve for the unknown.

Follow these four steps to solve an algebraic equation:

Step 1. Remove parentheses by multiplication.
Step 2. Combine separated unknowns.
Step 3. Do addition or subtraction first.
Step 4. Do multiplication or division last.

To remove parentheses, multiply each term inside the parentheses by the number outside the parentheses. If the parentheses are preceded by a negative sign or number, remove the parentheses by changing the sign of each term within the parentheses.

For example, $+4(x + 3)$ becomes $4 \cdot x + 4 \cdot 3 = 4x + 12$, and $-(3z + 2)$ becomes $-3z - 2$.

EXAMPLE 1. Solve for x in $4(x + 3) = 20$ — *Solve:* $4(x + 3) = 20$

Step 1. Remove parentheses by multiplication. — $4x + 12 = 20$

Step 2. Subtract 12 from 20. — $4x = 20 - 12$; $4x = 8$

Step 3. Divide 8 by 4. — $x = \frac{8}{4} = \mathbf{2}$

Answer: $\mathbf{x = 2}$

EXAMPLE 2. Solve for z in $4z - (3z + 2) = 5$ — *Solve:* $4z - (3z + 2) = 5$

Step 1. Remove parentheses by multiplication. — $4z - 3z - 2 = 5$

Step 2. Combine the z's. — $z - 2 = 5$

Step 3. Add 2 to 5. — $z = 5 + 2 = 7$

Answer: $z = 7$

Solve each equation below. Remember to remove the parentheses as your first step.

Example

$3(y - 6) = 15$
$3y - 18 = 15$
$3y = 15 + 18$
$3y = 33$
$y = \frac{33}{3}$
$\mathbf{y = 11}$

1. $2(a + 3) = 16$

2. $4(b - 2) = 8$

Example

$-x - 2(x - 4) = -1$
$-x - 2x + 8 = -1$
$-3x + 8 = -1$
$-3x = -1 - 8$
$-3x = -9$
$x = \frac{-9}{-3}$
$x = 3$

3. $2m - 3(m + 3) = -13$

4. $-3y + 2(2y - 1) = -6$

Parentheses may appear on both sides of an equation. Remove both sets of parentheses as your first step in solving for the unknown.

EXAMPLE: Solve for x in $3(x - 6) = 2(x + 3)$

Step 1. Remove both sets of parentheses by multiplication.

Step 2. Subtract $2x$ from $3x - 18$.

Step 3. Add 18 to 6.

Answer: $x = 24$

Check by substituting 24 for x.

Solve:

$3(x - 6) = 2(x + 3)$
$3x - 18 = 2x + 6$
$3x - 2x - 18 = 6$
$x - 18 = 6$
$x = 6 + 18$
$\mathbf{x = 24}$

Check:

$3(24 - 6) = 2(24 + 3)$
$3(18) = 2(27)$
$54 = 54$

Solve each equation below by first removing parentheses from each side.

Examples

$4(y - 3) = 3(y + 6)$
$4y - 12 = 3y + 18$
$4y - 3y - 12 = 18$
$y - 12 = 18$
$y = 18 + 12$
$y = 30$

$2(x + 4) = 8 - (x + 3)$
$2x + 8 = 8 - x - 3$
$2x + x + 8 = 8 - 3$
$3x + 8 = 5$
$3x = 5 - 8$
$3x = -3$
$x = \frac{-3}{3}$
$x = -1$

5. $5(z - 1) = 4(z + 4)$

6. $7(a - 2) = 6(a + 1)$

7. $3(y + 1) = 9 - (y + 2)$

8. $4(z - 2) = -2(z - 5)$

Solve each equation below. Check any problem you're unsure of on scratch paper.

9. $4(z + 2) = 2z + 40$

10. $3(y + 2) = 2(y + 5)$

11. $5(x - 3) = 4x - 7$

12. $3(\frac{x}{3} - 2) = 9$

13. $6(b - 2) = 12$

14. $5(x + 3) = 35$

15. $9 + 2(y - 4) = 13$

16. $4(a - 3) = 3 - (a + 5)$

17. $3z = 2(z - 3) + 9$

18. $8a - (a - 7) = 21$

19. $\frac{2}{3}(x + 9) = \frac{1}{3}(x + 6)$

20. $y + 2(y - 3) = 6$

21. $2(x + 5) = x + 4$

22. $4 - (5 - x) = 2(4 - x)$

23. $3(z - 2) = 2(z - 1)$

24. $7(x + 3) = 2(-6 - 2x)$

25. $5(y + 2) = 3y + 16$

26. $4(x + 6) = 2(x - 3)$

SOLVING AN EQUATION WITH PARENTHESES: APPLYING YOUR SKILLS

In the following examples, notice how parentheses are used to represent the information contained in the problems. Study these examples carefully before solving the word problems on the next page.

EXAMPLE 1. Three times the quantity a number minus 4 is equal to two times the sum of the number plus 3. What is the number?

Step 1. Let x = the unknown number
$3(x - 4)$ is three times the quantity x minus 4
$2(x + 3)$ is two times the sum of x plus 3

Step 2. Write an equation for the problem.

Solve: $3(x - 4) = 2(x + 3)$

Step 3. Solve the equation.

$$3(x - 4) = 2(x + 3)$$

a) Remove parentheses.

$$3x - 12 = 2x + 6$$

b) Subtract $2x$ from $3x - 12$.

$$3x - 2x - 12 = 6$$
$$x - 12 = 6$$

c) Add 12 to 6.

$$x = 6 + 12$$
$$x = 18$$

Answer: $\mathbf{x = 18}$

EXAMPLE 2. Mary, Anne, and Sally share living expenses. Anne pays $25 less rent than Mary. Sally pays twice as much rent as Anne. If the total rent is $365, how much rent does each pay?

Step 1. Let x = Mary's rent
$x - 25$ = Anne's rent
$2(x - 25)$ = Sally's rent

Hint: Since you know nothing about how much Mary pays for rent, let Mary's rent equal x.

Step 2. Write an equation for the problem.

Mary's + Anne's + Sally's = total rent.

$$x + (x - 25) + 2(x - 25) = 365$$

Step 3. Solve the equation.

a) Remove parentheses.

Solve: $x + (x - 25) + 2\ (x - 25) = 365$

b) Combine the x's and the numbers.

$$x + x - 25 + 2x - 50 = 365$$
$$4x - 75 = 365$$

c) Add 75 to 365.

$$4x = 365 + 75$$
$$4x = 440$$

d) Divide 440 by 4.

$$x = \frac{440}{4}$$
$$x = 110$$

Answer:

x = **$110, Mary's rent**
$x - 25$ = **$85, Anne's rent**
$2(x - 25)$ = **$170, Sally's rent**

Solve the following word problems.

1. Three times the sum of a number plus 1 is equal to two times the sum of the number plus 4. What is the number?

____

2. Five times the quantity x minus 1 is equal to three times the quantity x plus 9. What is x?

____

3. Four times a number minus 3 times the sum of the number plus 2 is equal to 5. What is the number?

____

4. Two-thirds times the quantity y minus 3 is equal to one-third y. What is y?

____

5. Maria, Amy, and Sadie share food expenses. Amy pays $10 a month less than Maria. Sadie pays twice as much as Amy. If the monthly food bill is $310, how much does each pay?

____

6. Frank, Sam, and Louis went to lunch. Sam's meal cost $.45 less than Frank's. Louis's meal cost twice as much as Sam's. If the bill came to $9.25, how much does each one owe?

____

7. Consecutive integers are whole numbers that follow one another, for example, 11, 12, and 13. If the sum of three consecutive integers is 54, what are the integers? (*Hint:* x = 1st integer, $x + 1$ = 2nd integer, $x + 2$ = 3rd integer.)

____

8. Consecutive even integers are even whole numbers that follow one another. Example: 12, 14, and 16. If the sum of three consecutive even integers is 66, what are the integers? (*Hint:* The first number is x, the second even integer is $x + 2$, the third is . . .)

____

RATIO

A *ratio* is a comparison of two numbers. For example, if there are five women and four men in a class, the ratio of women to men is 5 to 4.

The ratio 5 to 4 can be written in symbols in two ways:

(1) With a colon, 5 to 4 is written 5:4
(2) As a fraction, 5 to 4 is written $\frac{5}{4}$

In words, a ratio is always read with the word "to."

5:4 is read "Five to four."

A ratio, like a fraction, is always reduced to lowest terms. However, a ratio that is written as an improper fraction does not need to be changed to a mixed number.

EXAMPLE. Reduce the ratio 10 to 4 to lowest terms.

Step 1. Write the ratio 10 to 4 as a fraction.

10 to 4 = $\frac{10}{4}$

Step 2. Reduce the fraction $\frac{10}{4}$ to lowest terms.

$\frac{10}{4} = \frac{10 \div 2}{4 \div 2} = \frac{5}{2}$ *or* 5 to 2

Answer: **5 to 2** *or* **5:2**

Express each ratio in lowest terms.

Examples

8:6 = $\frac{8}{6} = \frac{4}{3}$

Answer: 4:3

6 to 10 = $\frac{6}{10} = \frac{3}{5}$

Answer: 3 to 5

1. 9:15 **2.** $\frac{15}{20}$ **3.** 100:28

4. 3 to 9 **5.** 12 to 8 **6.** 20 to 12

A whole number ratio is always written with a denominator of 1.

Example

8 to 2 = $\frac{8}{2} = \frac{4}{1}$

Answer: 4 to 1

7. 9 to 3 **8.** 20:4 **9.** $\frac{16}{8}$

RATIO: APPLYING YOUR SKILLS

Ratio problems are most often expressed in words. To solve a ratio word problem, be careful to write the ratio in the same order as it appears in the question.

EXAMPLE: Bill earns $800 a month. He pays $200 a month in rent. What is the ratio of his rent to his income?

Step 1. Express the ratio of rent to income. Since rent is mentioned first, it is the numerator of the ratio fraction.

rent to income $= \frac{\text{rent}}{\text{income}} = \frac{\$200}{\$800}$

Step 2. Reduce the ratio $\frac{\$200}{\$800}$ to lowest terms.

$\frac{\$200}{\$800} = \frac{\$200 \div 200}{\$800 \div 200} = \frac{1}{4}$

Answer: $\mathbf{\frac{1}{4}}$ *or* **1:4** *or* **1 to 4**

Note: If the question had asked for the ratio of income to rent, the answer would have been 4:1.

Solve the following problems using ratios.

1. A new sub-compact car gets 48 miles to the gallon in country driving and 32 miles to the gallon in the city. What is the ratio of country mileage to city mileage?

____

2. In a class with 30 students, there are 18 women and 12 men. What is the ratio of men to women in the class?

____

3. On an algebra quiz, Alice got 24 problems correct and 4 incorrect. What is the ratio of correct problems to the total number of problems on the quiz?

____

4. Find the ratio of 1 foot to 1 yard. (Change 1 yard to 3 feet before writing the ratio.)

____

5. What is the ratio of 18 inches to 1 yard? (Change 1 yard to 36 inches before writing the ratio.)

____

PROPORTION

A *proportion* is made up of two equal ratios. For example, $\frac{3}{4} = \frac{9}{12}$ is a proportion. A proportion can be written in symbols in two ways:

(1) With colons, 3:4 = 9:12
(2) As equal fractions, $\frac{3}{4} = \frac{9}{12}$

In words, a proportion is read as two equal ratios connected by the word "as."

3:4 = 9:12 is read "Three is to four as nine is to twelve."
$\frac{3}{4} = \frac{9}{12}$ is read "Three is to four as nine is to twelve."

In a proportion, the *cross-products* are equal. To find the cross-products, *cross multiply*. Multiply the numerator of each side of the proportion by the denominator of the other side.

In symbols, we can represent cross-multiplication by crossing arrows:

Cross Multiplication	**Equal Cross-Products**
$\frac{3}{4} \times \frac{9}{12}$	$3(12) = 4(9)$ $36 = 36$

Very often, a problem asks you to find a missing term in a proportion. To find the missing term, write the proportion as two equal fractions and represent the missing term by a letter. Use cross-multiplication to write an equation for the unknown and then solve.

EXAMPLE 1. Find the missing term in the proportion ____:8 = 4:16

Step 1. Represent the missing term by x, and write the proportion as two equal fractions. $\frac{x}{8} = \frac{4}{16}$

Step 2. Cross-multiply, and write equal cross-products. $\frac{x}{8} \times \frac{4}{16}$

$$16x = 8(4)$$
$$16x = 32$$

Step 3. Solve for x. Divide 32 by 16.

$$x = \frac{32}{16}$$
$$\mathbf{x = 2}$$

Answer: $\mathbf{x = 2}$

EXAMPLE 2. Find the missing term in the proportion $\frac{30}{20} = \frac{90}{y}$

Step 1. Cross-multiply, and write equal cross products. $\frac{30}{20} \times \frac{90}{y}$

$$30y = 20(90)$$
$$30y = 1800$$

Step 2. Solve for y. Divide 1800 by 30

$$y = \frac{1800}{30}$$
$$\mathbf{y = 60}$$

Answer: $\mathbf{y = 60}$

Write the following proportions as two equal fractions.

1. Six is to twelve as nine is to eighteen ______
2. Fourteen is to twenty as seven is to ten ______
3. Two is to five as thirty is to seventy-five ______
4. Ten is to two as five is to one ______
5. Nine is to three as twelve is to four ______

6. 5:10 = 20:40 **7.** 6:3 = 16:8 **8.** 3:4 = 15:20 **9.** 25:10 = 5:2

10. x:7 = 3:21 **11.** 5:y = 15:20 **12.** 6:4 = 12:z **13.** 30:24 = x:12

Find the missing term in each proportion.

Examples

___:9 = 1:3

$\frac{x}{9} \times \frac{1}{3}$

$3x = 9$

$x = \frac{9}{3}$

$\mathbf{x = 3}$

12:8 = 3:___

$\frac{12}{8} \times \frac{3}{x}$

$12x = 8(3)$

$12x = 24$

$x = \frac{24}{12}$

$\mathbf{x = 2}$

$\frac{y}{6} \times \frac{12}{18}$

$18y = 6(12)$

$18y = 72$

$y = \frac{72}{18}$

$\mathbf{y = 4}$

14. ___:4 = 6:8 **15.** ___:4 = 18:12 **16.** 1:___ = 4:20

17. 9:12 = 15:___ **18.** 14:7 = ___:5 **19.** 10:6 = 5:___

20. $\frac{15}{5} = \frac{x}{20}$ **21.** $\frac{7}{21} = \frac{9}{y}$ **22.** $\frac{z}{16} = \frac{200}{80}$

PROPORTION: APPLYING YOUR SKILLS

Writing a proportion is a useful method in solving many word problems. As in ratio problems, be very careful to write the terms of the proportion in the order stated in the problem.

EXAMPLE 1. A recipe calls for 2 cups of sugar for each 5 cups of flour. How many cups of sugar are needed for 15 cups of flour?

Step 1. Express each ratio of the proportion as cups of sugar to cups of flour. Let x represent the unknown number of cups of sugar.

In words, $\frac{\text{known cups of sugar}}{\text{known cups of flour}} = \frac{\textit{unknown}\text{ cups of sugar}}{\text{known cups of flour}}$

In symbols, $\frac{2}{5} = \frac{x}{15}$

Step 2. Solve the proportion by cross-multiplication.

$2(15) = 5x$

$30 = 5x$

$\frac{30}{5} = x$

$\mathbf{6} = x$

Answer: **6 cups of sugar** are needed for 15 cups of flour.

EXAMPLE 2. A light green paint is made by mixing 8 parts of white paint to 3 parts of green. How many quarts of green paint are mixed with 24 quarts of white to make the light green color?

Step 1. Express each ratio as amount of white to amount of green. Let x represent the unknown number of quarts of green paint.

In words, $\frac{\text{known parts of white}}{\text{known parts of green}} = \frac{\text{known quarts of white}}{\textit{unknown}\text{ quarts of green}}$

In symbols, $\frac{8}{3} = \frac{24}{x}$

Step 2. Solve the proportion by cross-multiplication.

$8x = 3(24)$

$8x = 72$

$x = \frac{72}{8}$

$x = \mathbf{9}$

Answer: **9 quarts of green paint** are mixed with 24 quarts of white.

Solve these problems by using proportions.

1. On a map of the United States, 2 inches represent 650 miles. How many inches of map distance represent 1,625 miles?

____

2. How much do 18 eggs cost when eggs are selling for 90 cents a dozen?

____

3. A car travels 450 miles in 8 hours. At this same rate, how long will it take to travel another 225 miles?

____

4. Georgia earns $35.40 in 8 hours. How much will she earn in 20 hours?

____

5. When he painted his house, Bob mixed 2 gallons of white paint with each 3 gallons of blue paint. How many gallons of white paint did he use if he used 6 gallons of blue?

____

6. If epoxy is to be mixed in a ratio of 3 parts hardener to 7 parts base, how many drops of hardener are required for 28 drops of base?

____

7. A new car went 102 miles on just 3 gallons of gas. How far can this car go on 18 gallons of gas?

____

8. If 8 ounces of steak costs $1.74, how much does 12 ounces of steak cost?

____

9. A light tan paint is made by mixing 5 parts of white paint to 2 parts of brown. How many quarts of brown must be mixed with 15 quarts of white to make the tan color?

____

10. If 12.7 centimeters are exactly 5 inches, how many centimeters are in 2 inches (to the nearest hundredth)?

____

SOLVING AN EQUATION WITH TWO UNKNOWNS

In our study of algebra, we have worked with equations containing one unknown. On these next few pages, we'll briefly look at equations that express a relationship between two unknowns. These equations will be used in the next chapter on rectangular coordinates.

For example, $y = 2x$ is an equation containing two unknowns, x and y.
In words, $y = 2x$ reads "y is equal to two times x."

An equation containing two unknowns has more than one solution. For each value of x that we choose, there is a matching value for y.

To solve an equation with two unknowns, follow these two steps:

Step 1. Choose several values for the second unknown.
Step 2. Substitute each value of the second unknown into the original equation and find a matching value for the first unknown.

To keep the solutions in neat order, write them in a Table of Values.

EXAMPLE 1. Solve for y in $y = 2x$, for $x = 0, 1, 2,$ and 3.
Step 1. Write the chosen values for x in the Table of Values.
Step 2. Substitute each value of x into the equation $y = 2x$.
Step 3. Fill in the Table of Values for y.

***x* value**	$y = 2x$
a) $x = 0$	$y = 2(0) = \mathbf{0}$
b) $x = 1$	$y = 2(1) = \mathbf{2}$
c) $x = 2$	$y = 2(2) = \mathbf{4}$
d) $x = 3$	$y = 2(3) = \mathbf{6}$

Table of Values

x	y
0	**0**
1	**2**
2	**4**
3	**6**

EXAMPLE 2. Solve for y in $y = 2x - 3$, for $x = -2, -1, 0,$ and 1.
Step 1. Write the chosen values for x in the Table of Values.
Step 2. Substitute each value of x into $y = 2x - 3$.
Step 3. Fill in the Table of Values for y.

***x* value**	$y = 2x - 3$
a) $x = -2$	$y = \overbrace{2(-2)}^{-4} - 3 = \mathbf{-7}$
b) $x = -1$	$y = \overbrace{2(-1)}^{-2} - 3 = \mathbf{-5}$
c) $x = 0$	$y = \overbrace{2(0)}^{0} - 3 = \mathbf{-3}$
d) $x = 1$	$y = \overbrace{2(1)}^{2} - 3 = \mathbf{-1}$

Table of Values

x	y
−2	**−7**
−1	**−5**
0	**−3**
1	**−1**

Complete each Table of Values by solving each equation for the given values of the second unknown.

1. $y = x + 4$

Table of Values

x	y
0	
1	
2	
3	

2. $y = 3x - 2$

Table of Values

x	y
1	
2	
3	
4	

3. $a = -2b + 3$

Table of Values

b	a
−1	
0	
1	
2	

4. $m = \frac{1}{2}n + 2$

Table of Values

n	m
0	
1	
2	
3	

5. $y = 2(x + 3)$

Table of Values

x	y
−2	
−1	
0	
1	

6. $y = -3(x - 1)$

Table of Values

x	y
−1	
0	
1	
2	

EQUATION SKILLS INVENTORY

1. Solve for x in $9 + x = 32$

2. Solve for y in $y - 17 = 8$

3. Solve for z in $9z = 162$

4. Solve for a in $\frac{a}{9} = 123$

5. Solve for b in $\frac{5}{11}b = 15$

6. Twelve times a certain number is equal to 96. What is the number?

7. Last year Jim paid $1,875 in taxes. If his tax bill is equal to $\frac{1}{6}$ of his total income, what was his total income last year?

8. Solve for x in $3x + 7 = 19$

9. Solve for y in $5 - \frac{y}{9} = 4$

10. Solve for z in $4z + 5z - 4 = 14$

11. Solve for x in $\frac{2}{3}x - \frac{1}{3}x = 12$

12. Solve for y in $4y + 6 = y + 18$

13. Solve for z in $3z - z - 5 = z + 8$

14. Mary and Lucy work together. Mary earns 7 dollars for every 5 dollars Lucy earns. If together they collect $156, how much does each earn?

15. Solve for x in $4(x - 3) = 3x - 4$

16. Solve for m in $5m - (m + 5) = 11$

17. Solve for y in $6(y + 3) = 4(y + 9)$

18. Alice, Joan, and Judy worked together to finish a painting job. Joan worked 8 hours less than Alice. Judy worked twice as long as Joan. If together they worked 72 hours on the job, how many hours did each work?

19. Reduce the ratio 18 to 26 to lowest terms.

20. There are 18 people in the class—12 men and 6 women. What is the ratio of men to women?

21. Solve for the missing term in the proportion $\frac{10}{x} = \frac{25}{15}$.

22. Jenny earns $26.00 during an 8-hour work day. At the same rate, how much money would Jenny earn working 14 hours?

23. Using two unknowns, write an equation that reads "y is equal to four times the quantity x plus nine."

24. Solve for y in $y = 4x - 7$, for $x = 2, 3, 4,$ and 5. Record your answers in the Table of Values at the right.

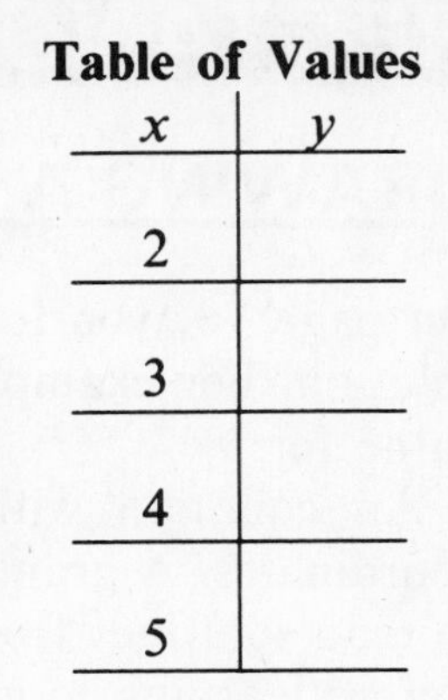

Table of Values

x	y
2	
3	
4	
5	

EQUATION SKILLS INVENTORY CHART

Circle the number of any problem that you missed and be sure to review the appropriate practice page. A passing score is 21 correct answers.

Problem Number	Skill Area	Practice Page
1	addition equation	47
2	subtraction equation	48
3	multiplication equation	49
4	division equation	50
5	fractional equation	51
6	applying equations	54
7	applying equations	54
8	equations: several operations	56
9	equations: several operations	56
10	equations: separated unknowns	58
11	equations: separated unknowns	58
12	equations: terms on both sides	62
13	equations: terms on both sides	62
14	applying equations	62
15	equations: parentheses	66
16	equations: parentheses	66
17	equations: parentheses	66
18	applying equations	70
19	ratio	72
20	applying ratio	73
21	proportion	74
22	applying proportion	76
23	equations: two unknowns	78
24	equations: two unknowns	78

If you missed more than 3 questions, you should review this chapter.

RECTANGULAR COORDINATES

DRAWING A GRAPH WITH RECTANGULAR COORDINATES

On page 78, you learned that an equation with two unknowns may have more than one solution. For example, in $y = 2x$, for each value of x you choose, there is a corresponding value for y.

An equation with two unknowns is often represented on a graph with rectangular coordinates. A graph has the advantage of displaying many solutions at once. You are able to read solutions from the graph without having to solve the equation to find each solution. Before learning to graph an equation, you will spend the next few pages building graphing skills.

A graph with rectangular coordinates is drawn by crossing a horizontal x number line and a vertical y number line. On the graph, the x number line is called the *x coordinate axis*, and the y number line is called the *y coordinate axis*. The point where the lines cross is called the *origin*.

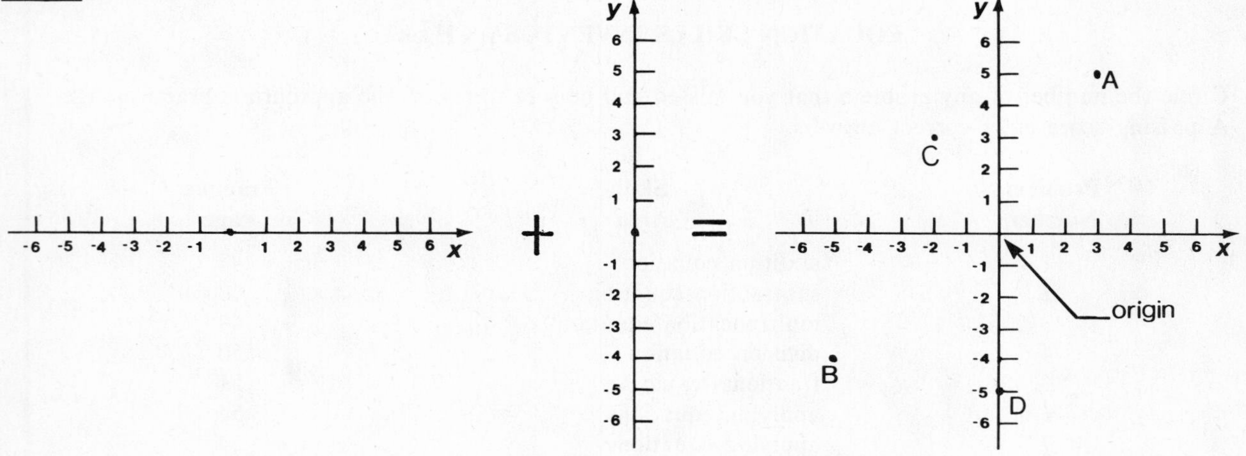

x number line + y number line = graph with rectangular coordinates

Each point on the graph is identified by two numbers: the *x coordinate* and the *y coordinate.*

The x coordinate tells how far the point is to the right or to the left of the y coordinate axis:

> A positive x coordinate indicates the point is to the right of the y coordinate axis, and a negative x coordinate indicates the point is to the left.

The y coordinate tells how far the point is above or below the x coordinate axis:

> A positive y coordinate indicates the point is above the x coordinate axis, and a negative y coordinate indicates the point is below.

To find the coordinates of a point on the graph, start at the point and follow these steps:

Step 1. To find the x coordinate, move directly up or down to a point on the x coordinate axis. This point is the x coordinate.

Step 2. To find the y coordinate, move to the right or left to a point on the y coordinate axis. This point is the y coordinate.

For simplicity, the x coordinate is often called the <u>*x value*</u> or just x; the y coordinate is often called the <u>*y value*</u> or just y.

To find the coordinates of the points in the following examples, use the graph on page 84.

EXAMPLE 1. Find the coordinates of Point A.

Step 1. Find the x coordinate: Start at Point A and move directly down to the x axis to the point +3. Point A is 3 units to the right of the y-axis.
x value = +3

Step 2. Find the y coordinate: Start at Point A and move directly across to the y axis to the point +5. Point A is 5 units above the x-axis.
y value = +5

EXAMPLE 2. Find the coordinates of Point B.

Step 1. Find the x coordinate: Start at Point B and move directly up to the x-axis to the point -5. Point B is 5 units to the left of the y-axis.
x value = –5

Step 2. Find the y coordinate: Start at Point B and move directly across to the y-axis to the point -4. Point B is 4 units below the x-axis.
y value = –4

EXAMPLE 3. The coordinates of Point C are: **$x = -2$ $y = +3$**

EXAMPLE 4. The coordinates of Point D are: **$x = 0$ $y = -5$**

EXAMPLE 5. The coordinates of the origin are: **$x = 0$ $y = 0$**

Find the coordinates of each graphed point below.

1.

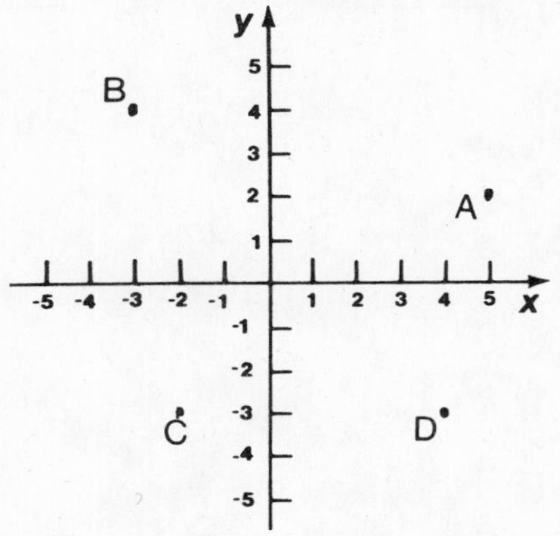

Point A: x coordinate =
y coordinate =

Point B: x coordinate =
y coordinate =

Point C: x value =
y value =

Point D: x =
y =

WRITING COORDINATES AS AN ORDERED PAIR

An *ordered pair* is made up of two numbers written in a specific order within parentheses. For example $(+3,-2)$ is an ordered pair.

In the study of algebra, coordinates of a point on a graph are usually written as an ordered pair. The x coordinate is written first within the parentheses, followed by a comma and the y coordinate. In symbols, you write (x, y).

EXAMPLE 1. Write the following coordinates as an ordered pair.
$x = -4$
$y = +5$

Step 1. Write the x coordinate as the first number inside a set of parentheses and follow the number with a comma.
$(-4,\quad)$

Step 2. Write the y coordinate second, following the comma.
$\mathbf{(-4,+5)}$

EXAMPLE 2. Write the x coordinate and the y coordinate of the ordered pair. $(3\frac{1}{2},-2)$.

Step 1. Identify the x coordinate as the first number of the pair.
$\boldsymbol{x = 3\frac{1}{2}}$

Step 2. Identify the y coordinate as the second number of the pair.
$\boldsymbol{y = -2}$

Write each pair of coordinates as an ordered pair.

1. $x = +3$, $y = +5$ ______ $x = -4$, $y = -2$ ______ $x = 7$, $y = 0$ ______ $x = 0$, $y = -3$ ______

2. $x = +\frac{2}{3}$, $y = -7$ ______ $x = 6$, $y = 1$ ______ $x = -3$, $y = \frac{1}{2}$ ______ $x = -\frac{3}{4}$, $y = -2\frac{1}{4}$ ______

Identify the x and y coordinates in each ordered pair below.

3. $(+2,+6)$ $x =$ $y =$ $(-3,+4)$ $x =$ $y =$ $(-7,-5)$ $x =$ $y =$ $(\frac{1}{2},+5)$ $x =$ $y =$

4. $(-3,2)$ $x =$ $y =$ $(\frac{2}{3},-4)$ $x =$ $y =$ $(4,7)$ $x =$ $y =$ $(-\frac{2}{3},-1)$ $x =$ $y =$

Write the coordinates of each graphed point below as an ordered pair.

5. **6.**

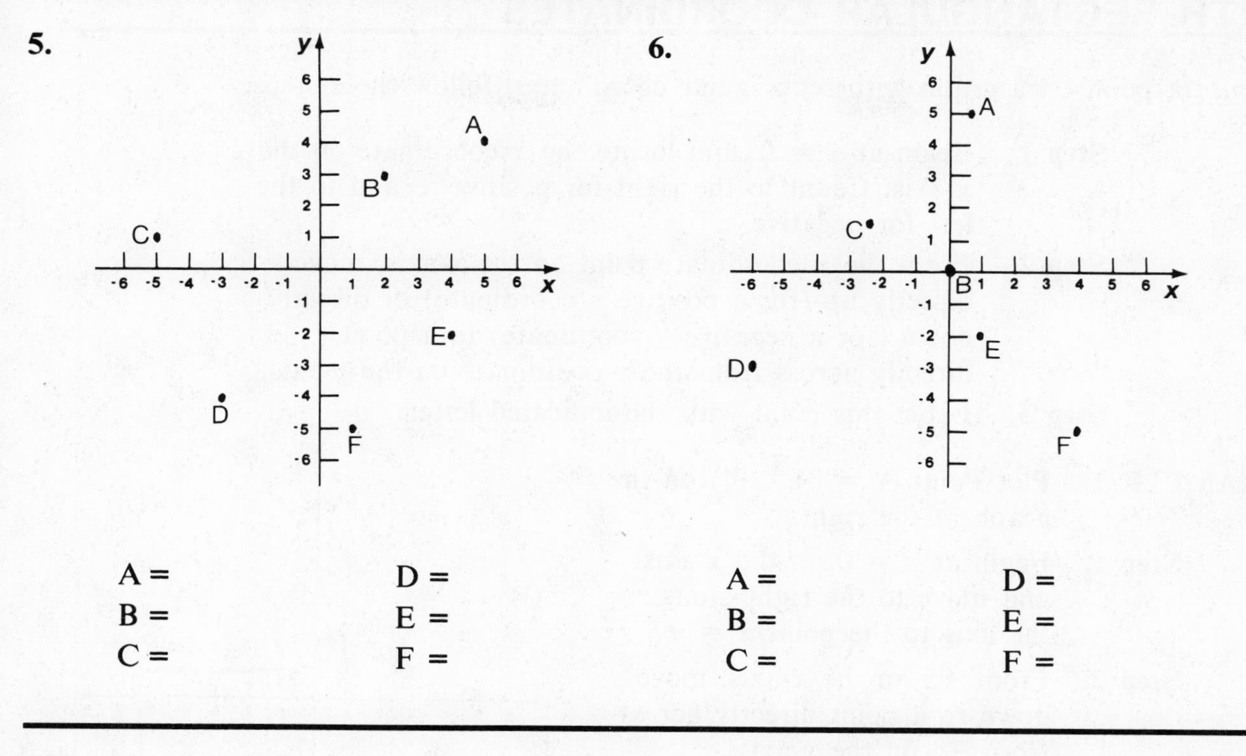

A =
B =
C =
D =
E =
F =

A =
B =
C =
D =
E =
F =

Points on a graph may stand alone or they may be connected. A common example of connected points is the *graphed line*. A graphed line is a straight line drawn through a set of plotted points. You read coordinates of points on a graphed line in the same way that you read coordinates of points standing alone.

---

7. **8.**

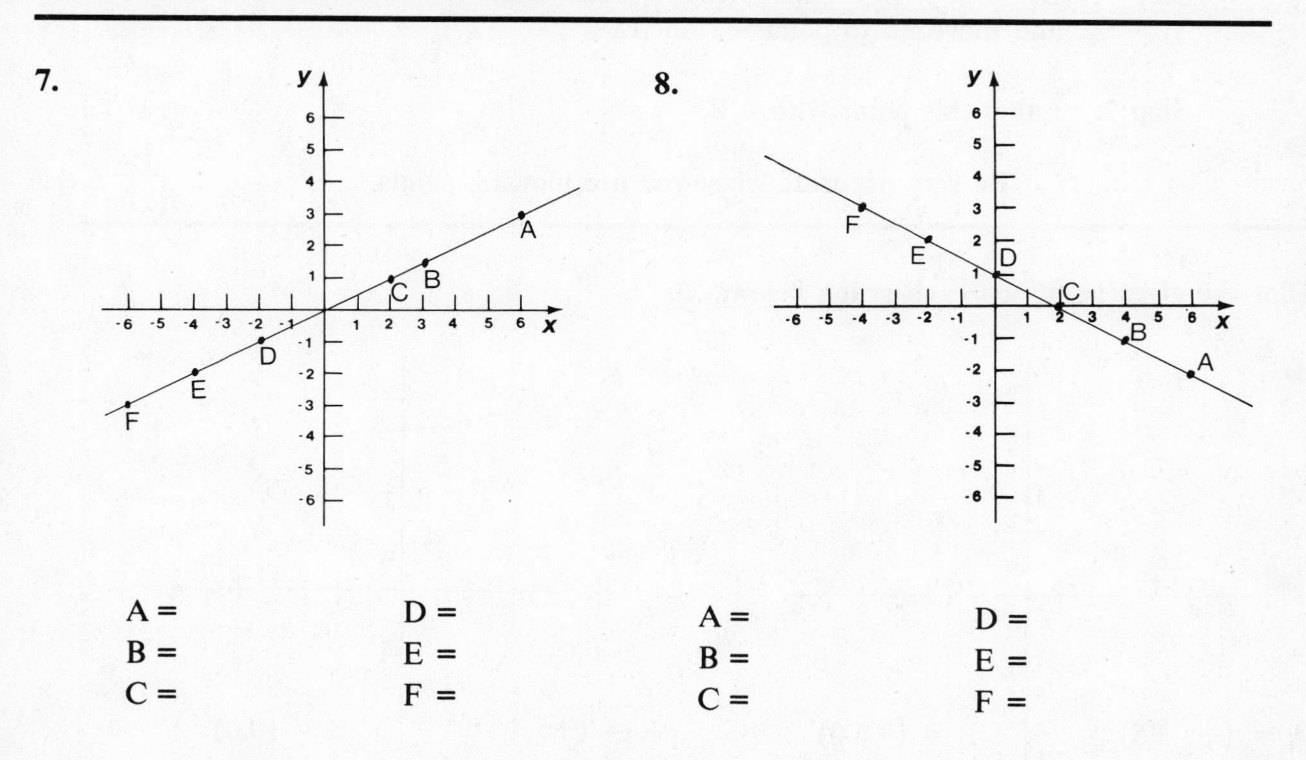

A =
B =
C =
D =
E =
F =

A =
B =
C =
D =
E =
F =

PLOTTING POINTS ON A GRAPH WITH RECTANGULAR COORDINATES

To *plot* a point on a graph with rectangular coordinates, follow these steps:

Step 1. Begin at $x = 0$, and locate the x coordinate on the x-axis. Count to the right for positive x and to the left for negative x.

Step 2. From the x coordinate point on the x-axis, move directly up (for a positive y coordinate) or directly down (for a negative y coordinate) to a point directly across from the y coordinate on the y-axis.

Step 3. Label this point with the indicated letter.

EXAMPLE 1. Plot Point A = (+5,−4) on the graph at the right.

Step 1. Begin at $x = 0$ on the x-axis, and move to the right along the axis to the point $x = +5$.

Step 2. From +5 on the x-axis, move down to a point directly across from −4 on the y-axis.

Step 3. Label this point with an **A**.

EXAMPLE 2. Plot Point B = (0,+3)

Step 1. Begin at $x = 0$ (this is the x coordinate) and move up to point +3 on the y-axis.

Step 2. Label this point with a **B**.

Be very accurate when you are plotting points.

Plot the given points on each graph below.

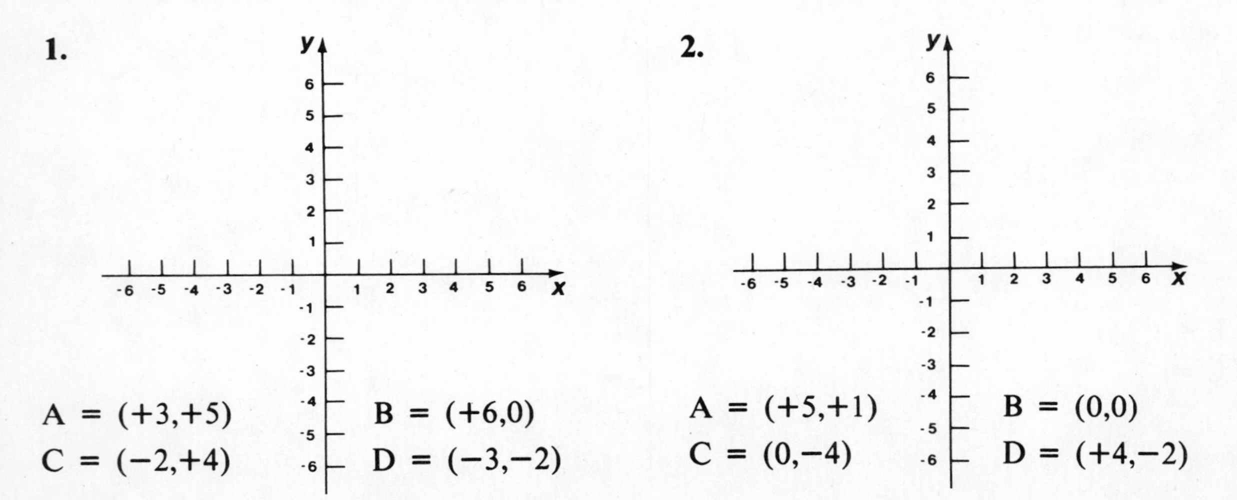

1. A = (+3,+5) B = (+6,0) C = (−2,+4) D = (−3,−2)

2. A = (+5,+1) B = (0,0) C = (0,−4) D = (+4,−2)

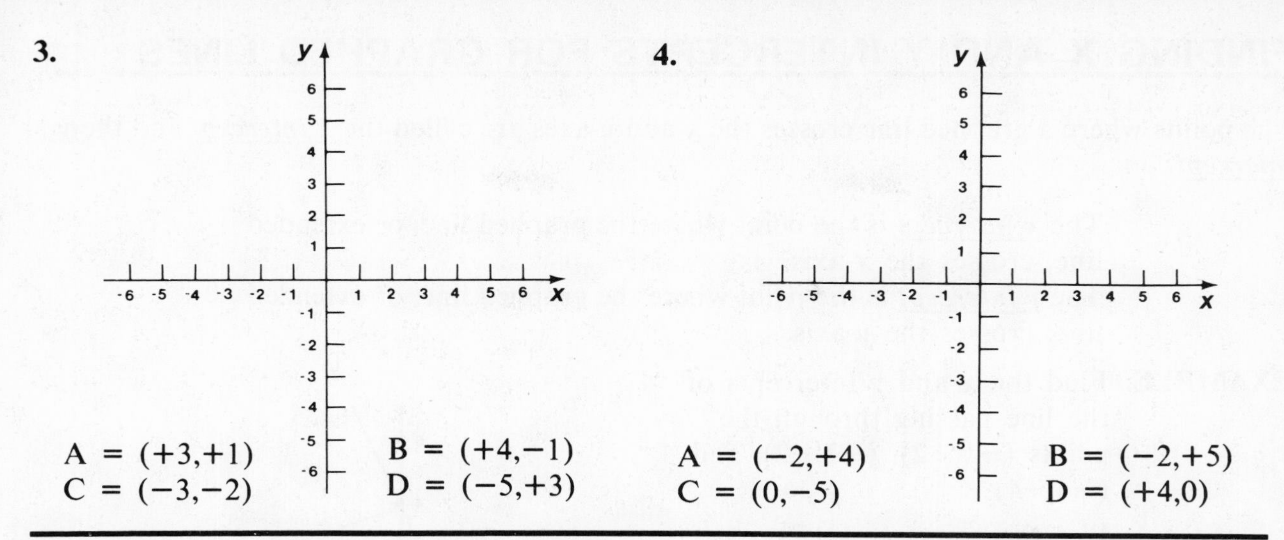

3.

A = (+3,+1) B = (+4,−1)
C = (−3,−2) D = (−5,+3)

4.

A = (−2,+4) B = (−2,+5)
C = (0,−5) D = (+4,0)

As you've seen, plotted points may lie on a straight line. To draw a <u>*graphed line*</u>, draw a straight line through these points and extend this line to the edges of the graph.

EXAMPLE: Plot the points (+1,+2), (+2,+4), and (−2,−4). Connect these points by a line extending to the edges of the graph.

Step 1. Plot the points. Notice that a point may be labeled simply by writing its coordinates next to the plotted point.

Step 2. Use a ruler to connect the plotted points and extend the line to the edges of the graph.

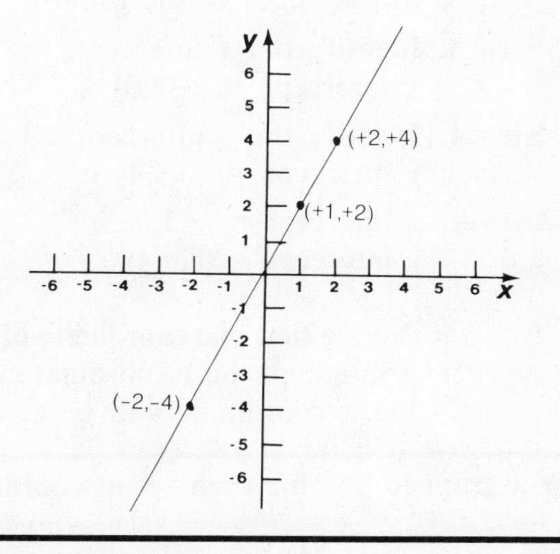

Draw the graphed line that passes through each set of points below.

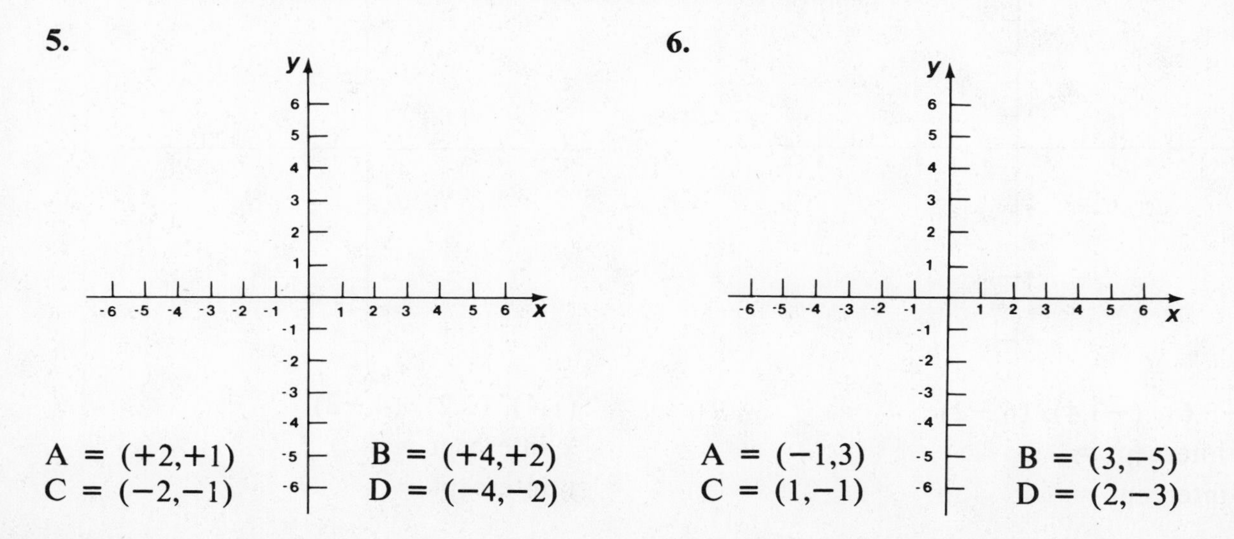

5.

A = (+2,+1) B = (+4,+2)
C = (−2,−1) D = (−4,−2)

6.

A = (−1,3) B = (3,−5)
C = (1,−1) D = (2,−3)

FINDING X AND Y INTERCEPTS FOR GRAPHED LINES

The points where a graphed line crosses the x and y-axes are called the *x intercept* and the *y intercept*.

> The *x intercept* is the point where the graphed line, or extended line, crosses the x-axis.
> The *y intercept* is the point where the graphed line, or extended line, crosses the y-axis.

EXAMPLE: Find the x and y intercepts of the line passing through the points $(-1,+2)$, $(-3,-2)$, and $(-4,-4)$.

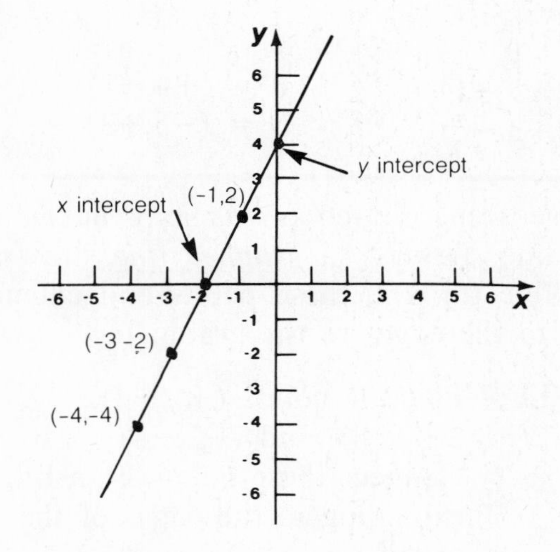

Step 1. Plot the points $(-1,+2)$, $(-3,-2)$, and $(-4,-4)$

Step 2. Draw a line connecting the three points and extend the line to the edges of the graph.

Step 3. Identify the x intercept:
x intercept $= (-2,0)$

Step 4. Identify the y intercept:
y intercept $= (0,+4)$

Answer: **x intercept $= (-2,0)$**
y intercept $= (0,+4)$

Notice that one coordinate of each intercept will be zero. For the x intercept the y coordinate will be zero, and for the y intercept the x coordinate will be zero.

Draw a graphed line for each set of coordinates and find the x and y intercepts of the line.

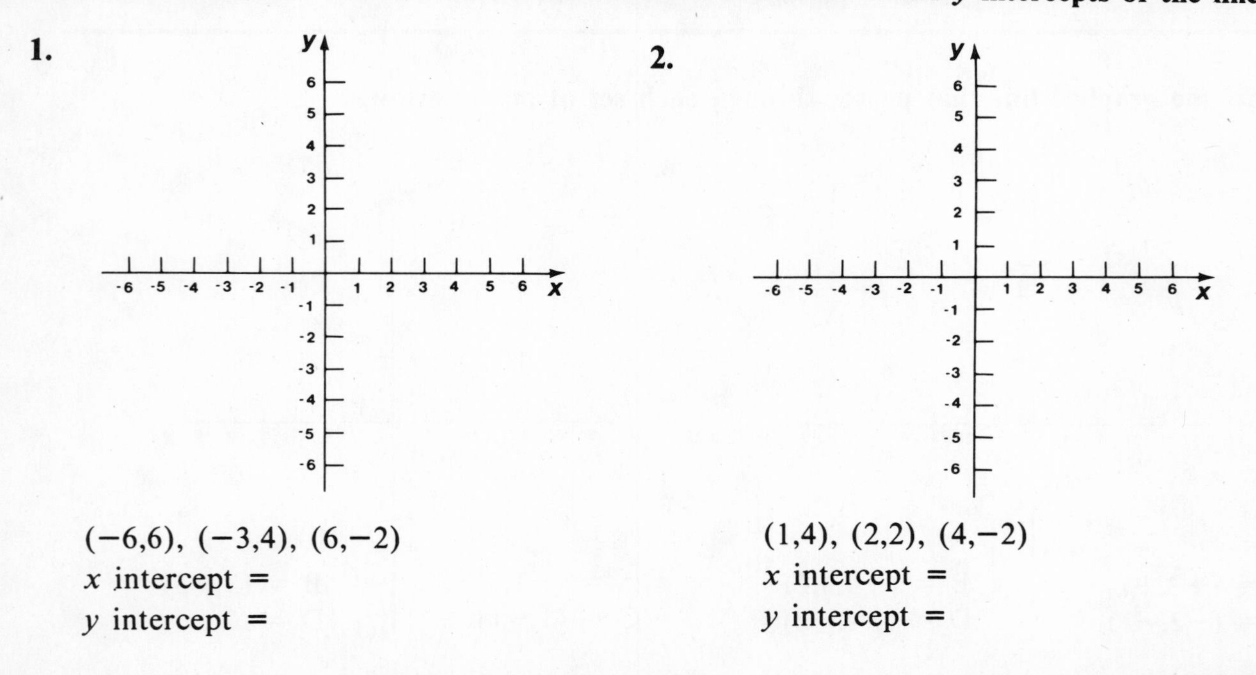

1. $(-6,6)$, $(-3,4)$, $(6,-2)$
x intercept =
y intercept =

2. $(1,4)$, $(2,2)$, $(4,-2)$
x intercept =
y intercept =

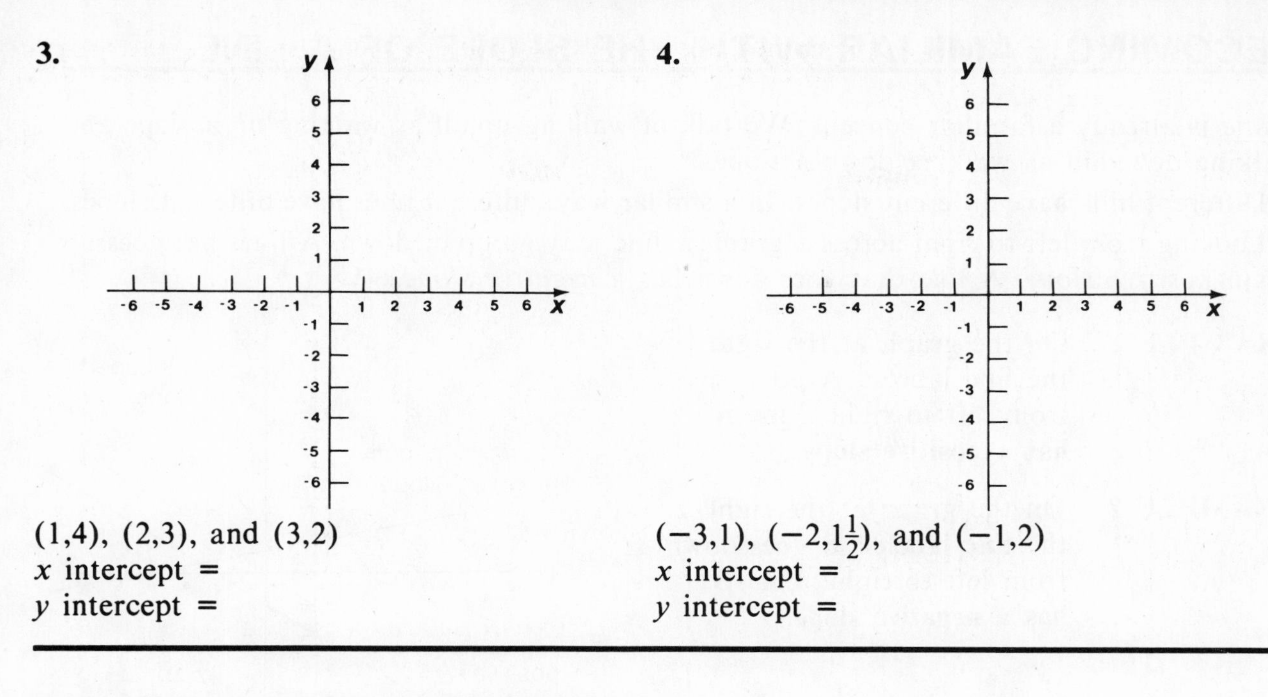

3. (1,4), (2,3), and (3,2)
x intercept =
y intercept =

4. (−3,1), $(-2,1\frac{1}{2})$, and (−1,2)
x intercept =
y intercept =

The previous exercises showed graphing by coordinates. A straight line can also be graphed if its x and y intercepts are known.

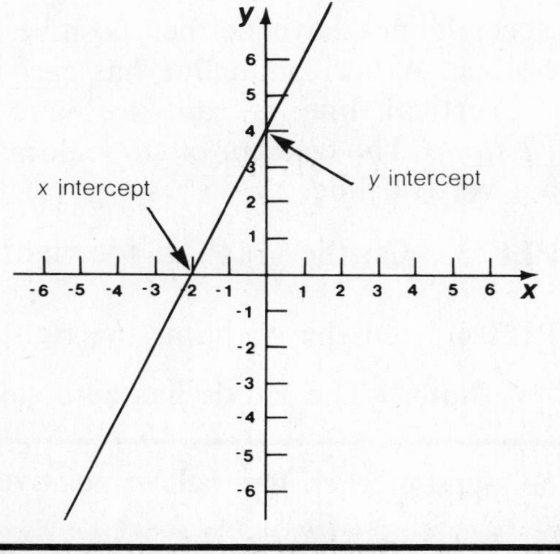

EXAMPLE: Graph the line described by
x intercept = (−2,0)
y intercept = (0,4).

Step 1. Mark the point (−2,0).
Step 2. Mark the point (0,4).
Step 3. Draw a straight line through the marked x and y intercepts.

Graph the lines that have the given x and y intercepts below.

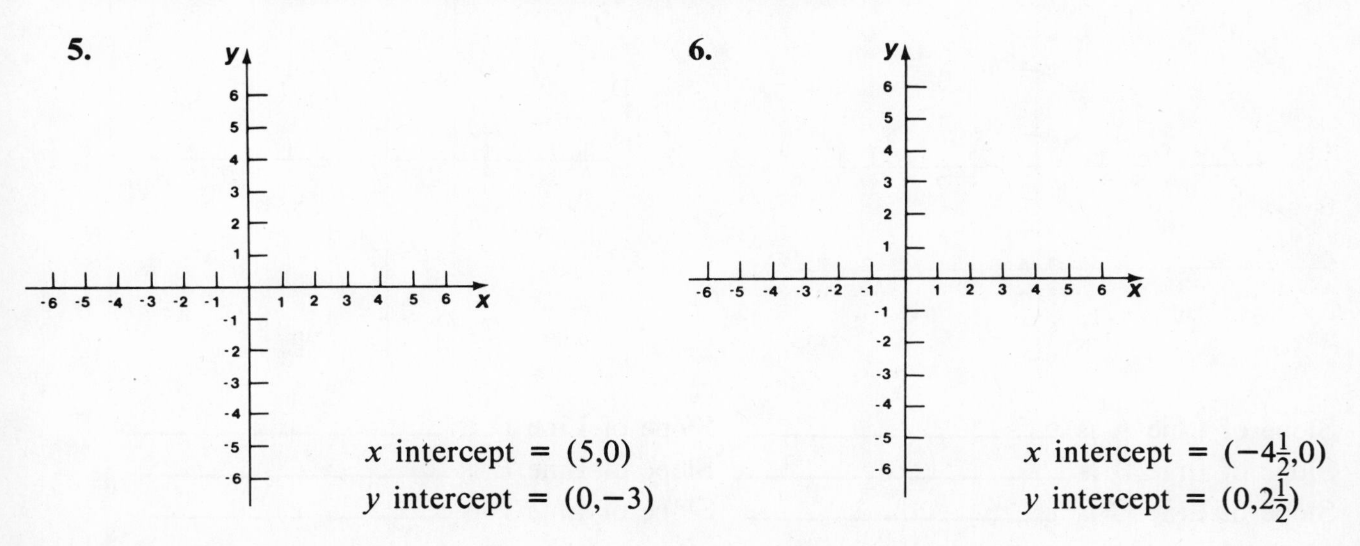

5. x intercept = (5,0)
y intercept = (0,−3)

6. x intercept = $(-4\frac{1}{2},0)$
y intercept = $(0,2\frac{1}{2})$

BECOMING FAMILIAR WITH THE SLOPE OF A LINE

Slope is already a familiar concept. We talk of walking uphill as walking up a slope and walking downhill as walking down a slope.

Different hills have different slopes. In a similar way, different lines have different slopes.

Looking from left to right across a graph, a line may go up or down. A line that goes up has a *positive slope*. A line that goes down has a *negative slope*.

EXAMPLE 1. On the graph at the right, the line labeled A goes up from left to right. **Line A has a positive slope.**

EXAMPLE 2. On the graph at the right, the line labeled B goes down from left to right. **Line B has a negative slope.**

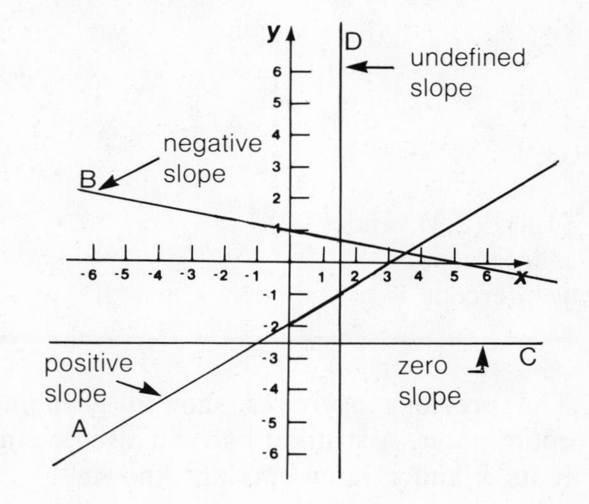

Two special lines have neither positive nor negative slope. A horizontal line has *zero (0) slope*. A vertical line is said to have an *undefined slope*. The concept of slope does not apply to a vertical line.

EXAMPLE 3. On the graph at the right, the line labeled C **has zero slope.**

EXAMPLE 4. On the graph at the right, the line labeled **D has an undefined slope.**

Note: The x-axis has zero slope. The y-axis has undefined slope.

Label the slope of each line below: positive, negative, zero, or undefined.

1. **2.**

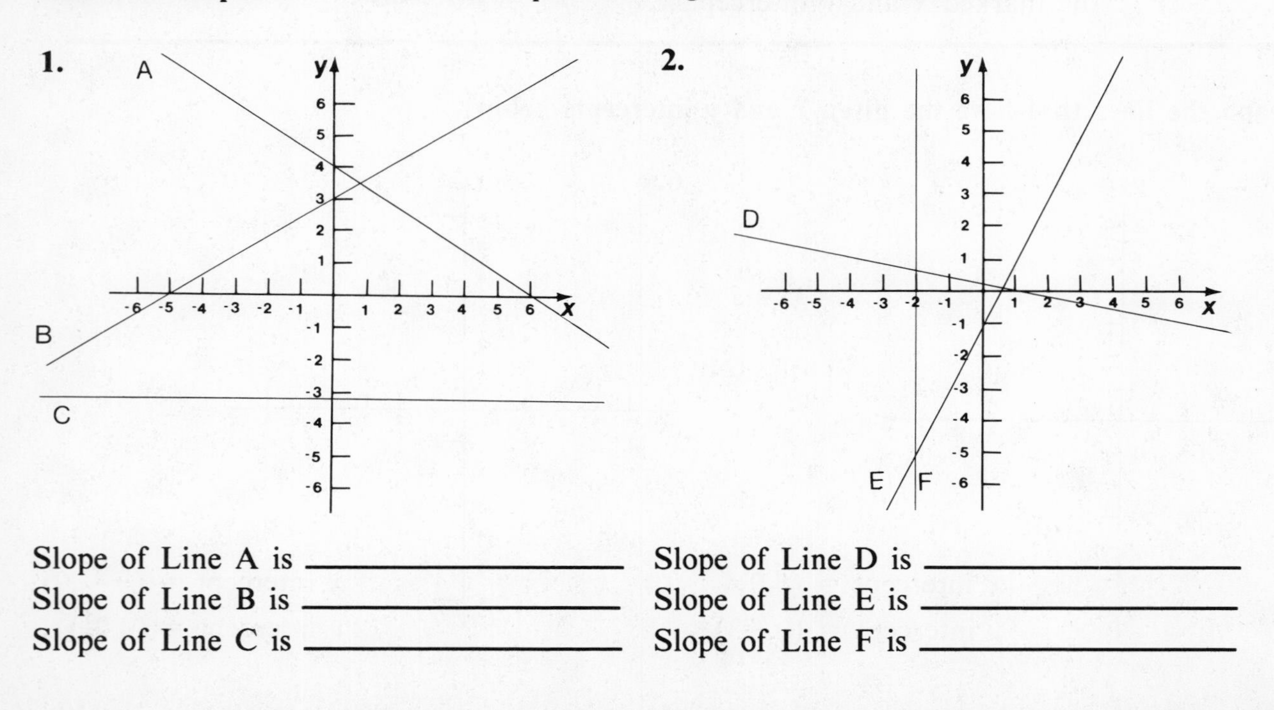

Slope of Line A is ______________

Slope of Line B is ______________

Slope of Line C is ______________

Slope of Line D is ______________

Slope of Line E is ______________

Slope of Line F is ______________

FIGURING OUT THE SLOPE

Besides knowing whether the slope of a line is positive or negative, you can figure out its exact measurement. To do this, choose two points that are on the line and then follow these three steps:

Step 1. Subtract the y coordinate of the second point from the y coordinate of the first point.
Step 2. Subtract the x coordinate of the second point from the x coordinate of the first point.
Step 3. Divide your answer from Step 1 by your answer in Step 2.

EXAMPLE: What is the slope of a line with points (1,2) and (3,6)? (It is possible to solve this without a picture of a graph.)

Step 1. Subtract the y coordinates $6-2=4$
Step 2. Subtract the x coordinates $3-1=2$
Step 3. Divide the answer from Step 1 by the answer in Step 2. $\frac{4}{2}=\mathbf{2}$

Answer: The slope of the line is 2.

Remember that you can have a positive or negative slope; your answer may be either positive or negative. If you are not sure of the sign, review the lesson on dividing signed numbers. The slope may also be a proper or an improper fraction.

Figure out the slope of the line having the following points.

1. (5,7) and (1,3)

2. (7,6) and (1,2)

3. (4,5) and (2,6)

4. (3,8) and (6,2)

5. (6,0) and (0,4)

GRAPHING A LINEAR EQUATION

The equation $y = 2x - 3$ is an example of a *linear equation*. When solutions of a linear equation are plotted on a graph, they always lie on a straight line. Drawing this line of solutions is called *graphing the equation.*

To graph a linear equation, follow these steps:

Step 1. Choose three values for x. For each value of x, find the corresponding value for y.

Step 2. Plot the three points on the graph and connect them with a line extending to the edges of the graph.

Note: Only two points are needed to graph a line, but the third point serves as a check that you plotted the first two points correctly.

EXAMPLE: Graph the equation $y = 2x - 3$.

Step 1. Let $x = 0$, 1, and 2. Solve the equation $y = 2x - 3$ for these values of x. Make a Table of Values.

x value	$y = 2x - 3$
$x = 0$	$y = 2(0) - 3 = -3$
$x = +1$	$y = 2(1) - 3 = -1$
$x = +2$	$y = 2(2) - 3 = +1$

x	y
0	−3
+1	−1
+2	+1

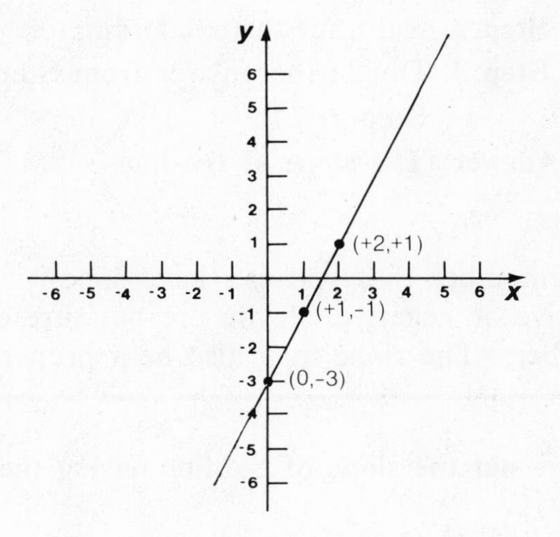

Step 2. Write the coordinates in the Table of Values as ordered pairs to be plotted: $(0,-3)$, $(+1,-1)$, and $(+2,+1)$

Step 3. Plot the points, and connect them with a straight line extending to the edges of the graph.

In this book we have assigned values of x for you to work with. In your future work with algebra, you may be asked to choose values for one of the unknowns.

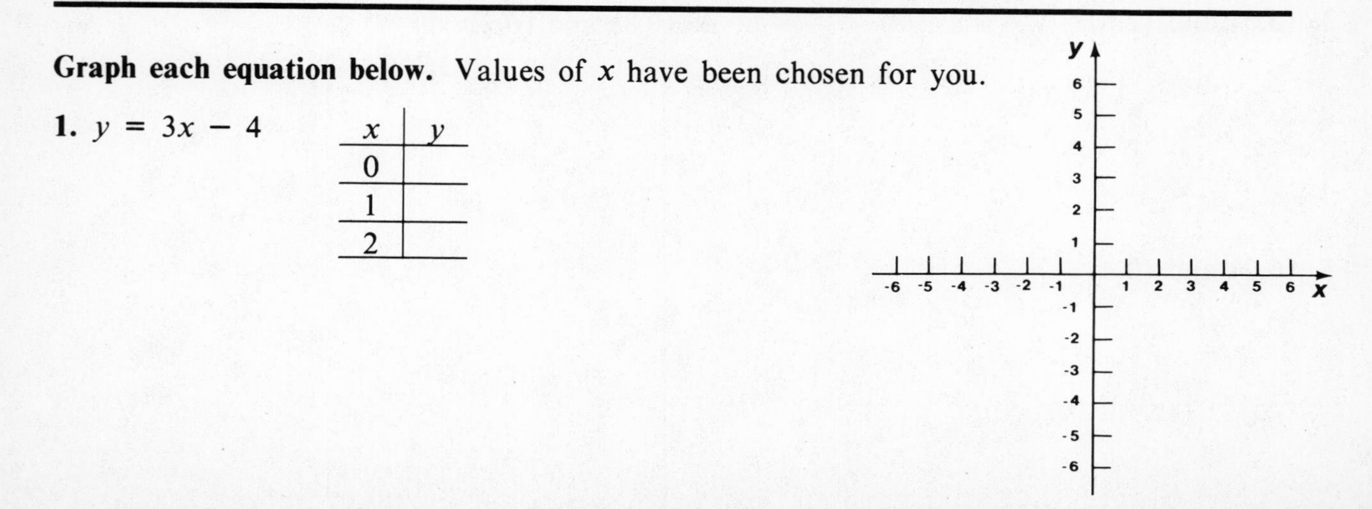

Graph each equation below. Values of x have been chosen for you.

1. $y = 3x - 4$

x	y
0	
1	
2	

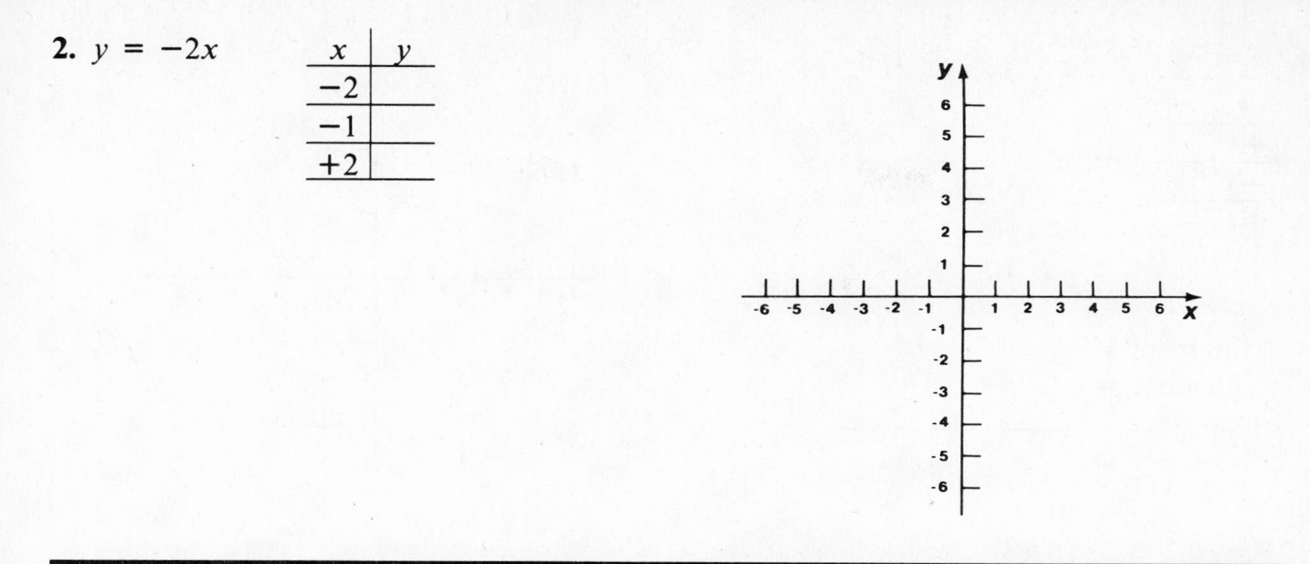

2. $y = -2x$

x	y
-2	
-1	
$+2$	

To find x and y intercepts and to name the slope of a graphed linear equation, follow the steps given in the example.

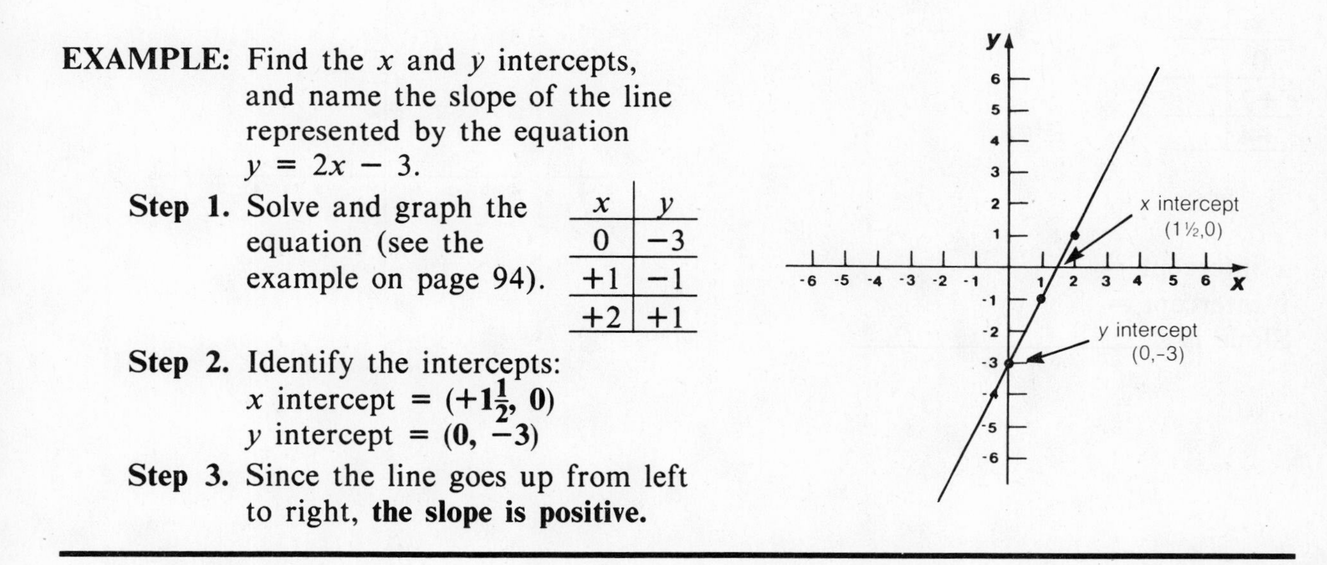

EXAMPLE: Find the x and y intercepts, and name the slope of the line represented by the equation $y = 2x - 3$.

Step 1. Solve and graph the equation (see the example on page 94).

x	y
0	-3
$+1$	-1
$+2$	$+1$

Step 2. Identify the intercepts:
x intercept = $(+1\frac{1}{2}, 0)$
y intercept = $(0, -3)$

Step 3. Since the line goes up from left to right, **the slope is positive.**

Graph each linear equation below. Find the x and y intercepts of each line and determine whether the slope is positive or negative.

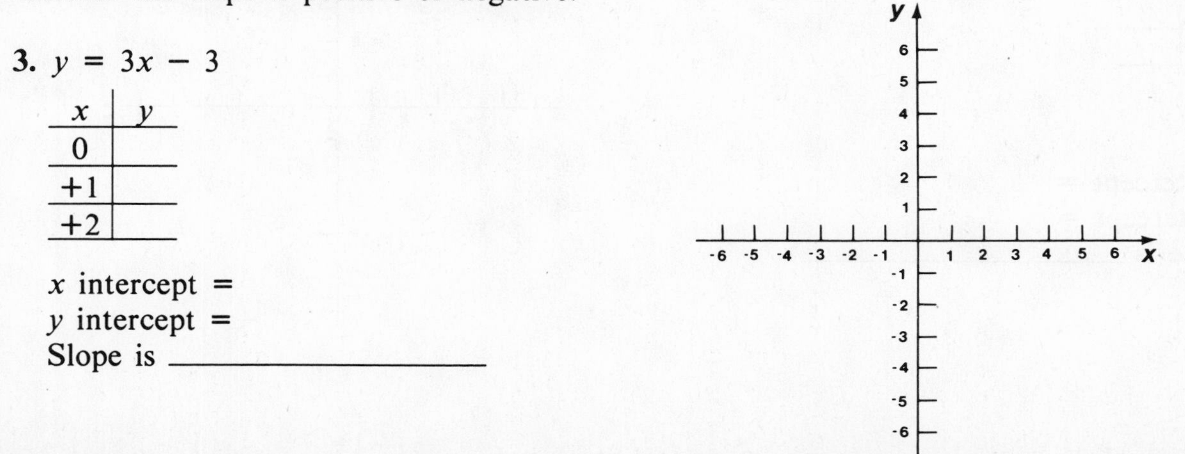

3. $y = 3x - 3$

x	y
0	
$+1$	
$+2$	

x intercept =
y intercept =
Slope is ____________________

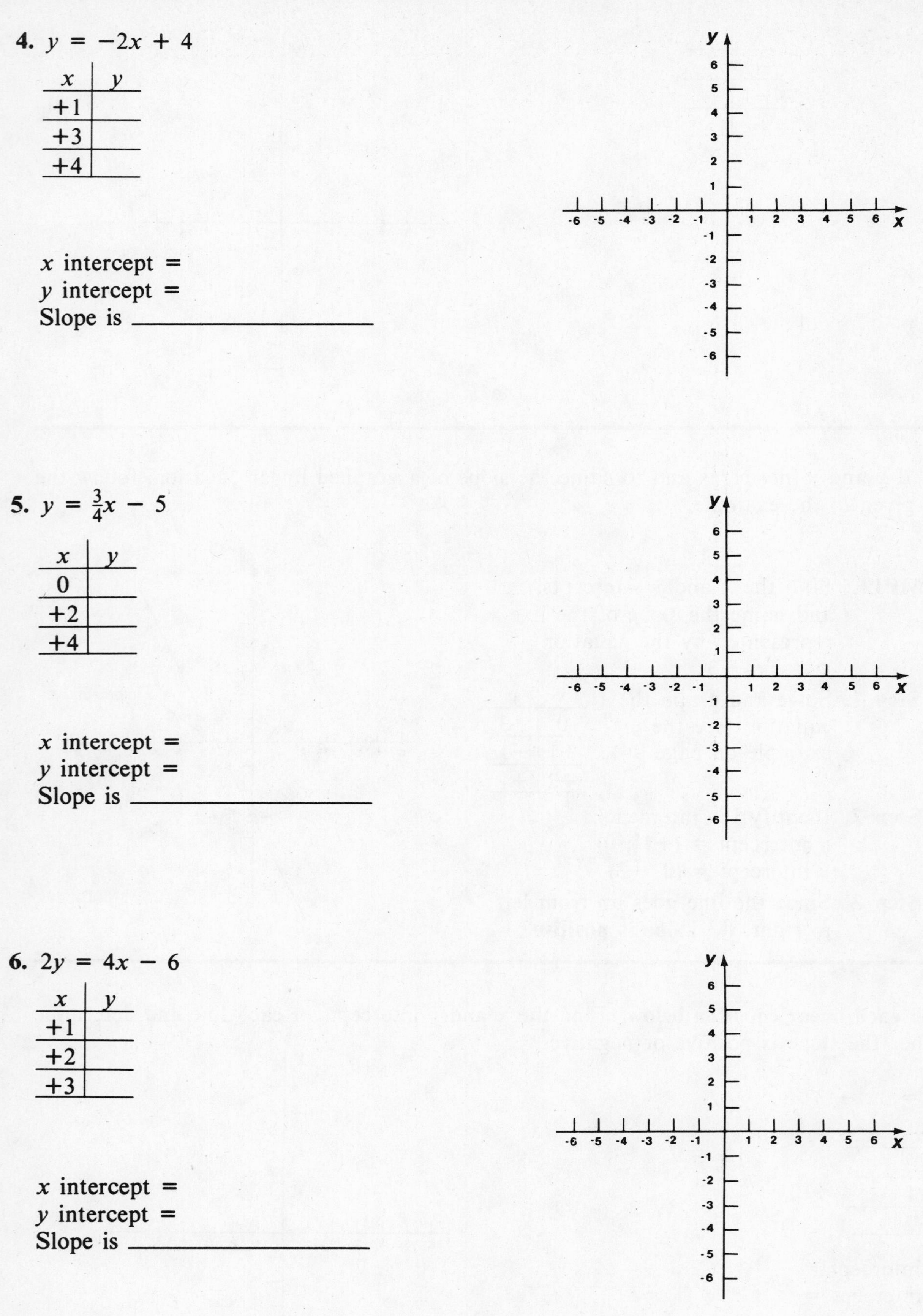

4. $y = -2x + 4$

x	y
+1	
+3	
+4	

x intercept =
y intercept =
Slope is ____________________

5. $y = \frac{3}{4}x - 5$

x	y
0	
+2	
+4	

x intercept =
y intercept =
Slope is ____________________

6. $2y = 4x - 6$

x	y
+1	
+2	
+3	

x intercept =
y intercept =
Slope is ____________________

RECTANGULAR COORDINATES SKILLS INVENTORY

1. On the graph at the right, label the following as indicated:
   a) the x-axis with an x
   b) the y-axis with a y
   c) the origin with (0,0)
   d) the x coordinates from -5 to $+5$
   e) the y coordinates from -5 to $+5$

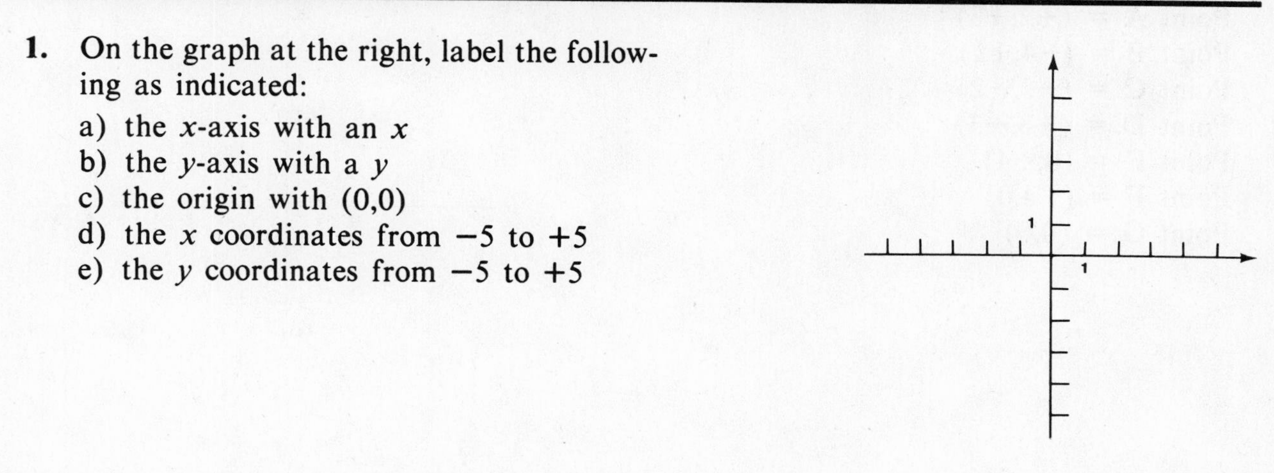

2. Write the following coordinates as an ordered pair.
   x coordinate $= -6$
   y coordinate $= -3$

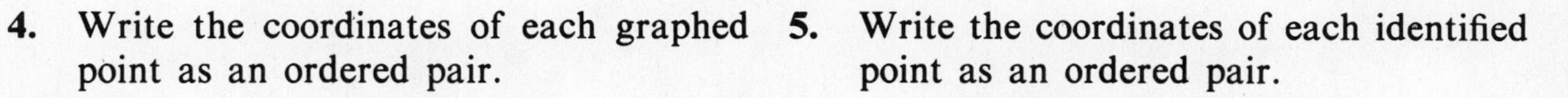

3. Identify the x and y coordinates in the ordered pair $(-4,+5)$.
   x coordinate =
   y coordinate =

4. Write the coordinates of each graphed point as an ordered pair.

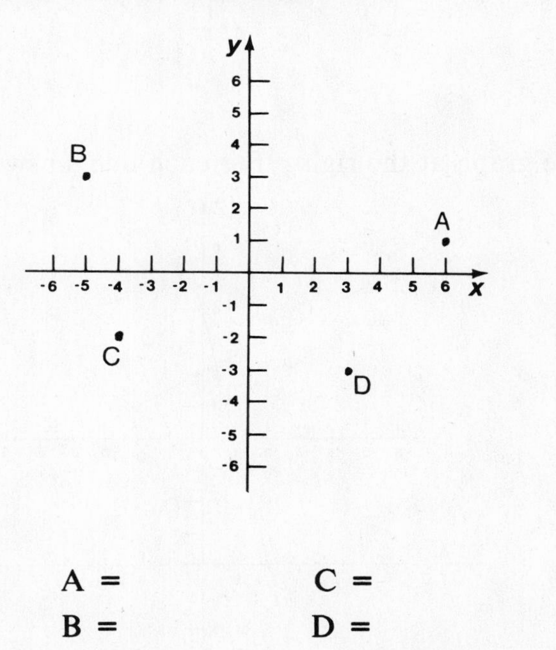

A =
B =
C =
D =

5. Write the coordinates of each identified point as an ordered pair.

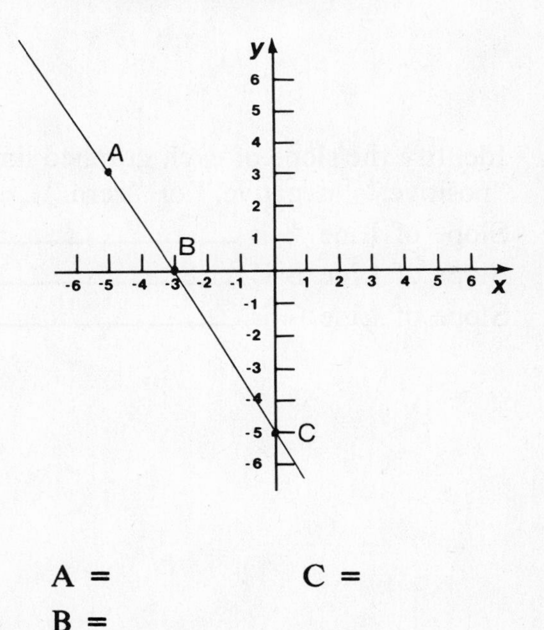

A =
B =
C =

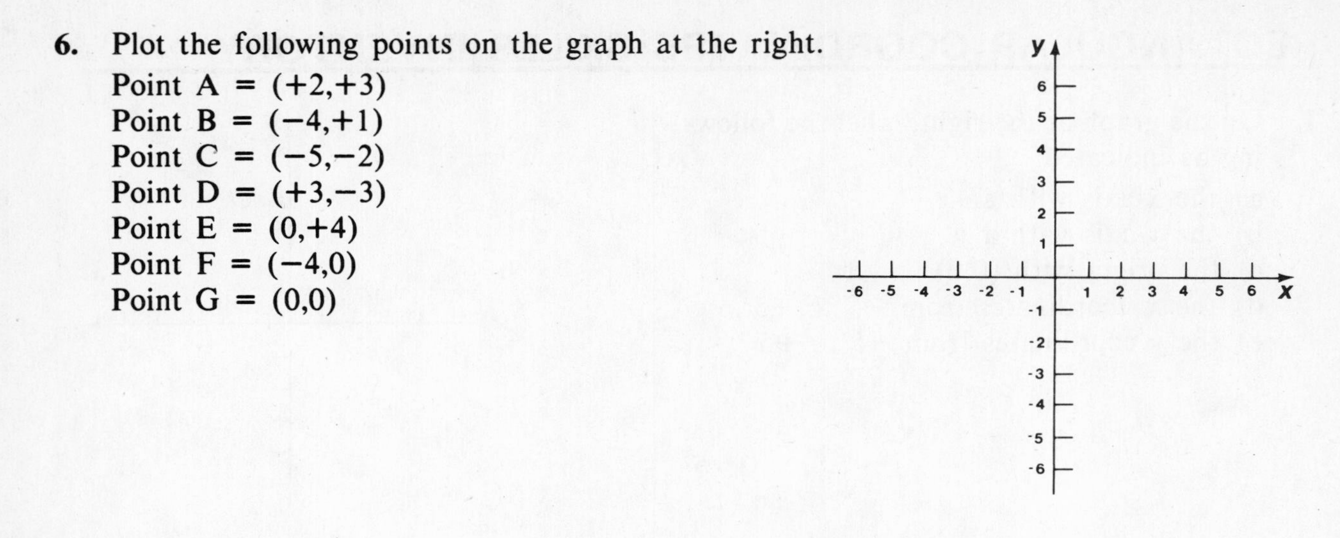

6. Plot the following points on the graph at the right.

Point A = (+2,+3)
Point B = (−4,+1)
Point C = (−5,−2)
Point D = (+3,−3)
Point E = (0,+4)
Point F = (−4,0)
Point G = (0,0)

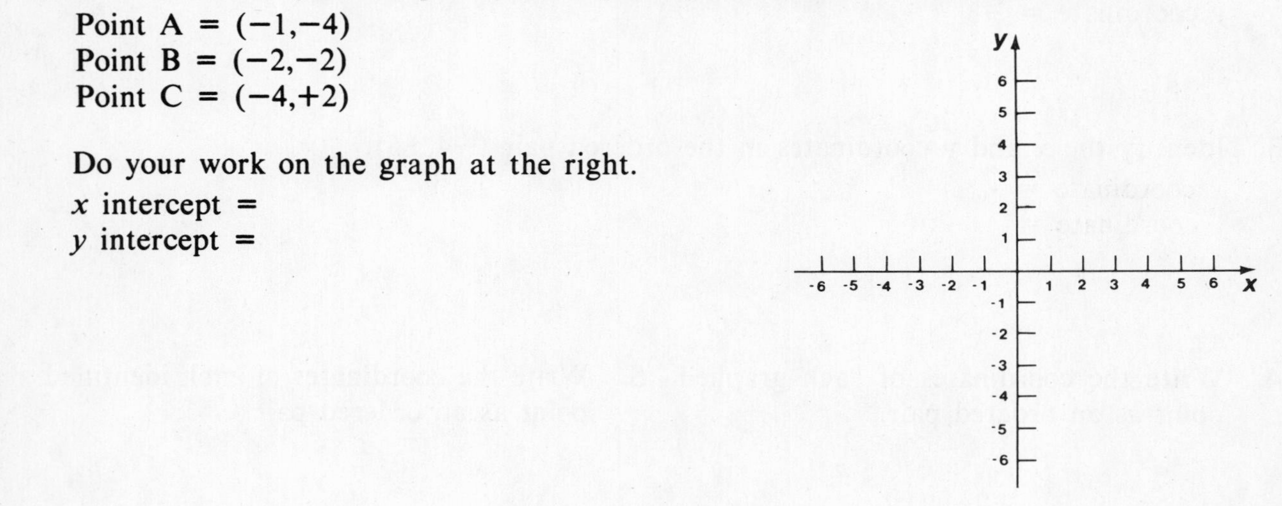

7. Find the x and y intercepts of a line passing through the following points:

Point A = (−1,−4)
Point B = (−2,−2)
Point C = (−4,+2)

Do your work on the graph at the right.

x intercept =
y intercept =

8. Identify the slope of each graphed line on the graph at the right. For each line, answer "positive," "negative," or "zero."

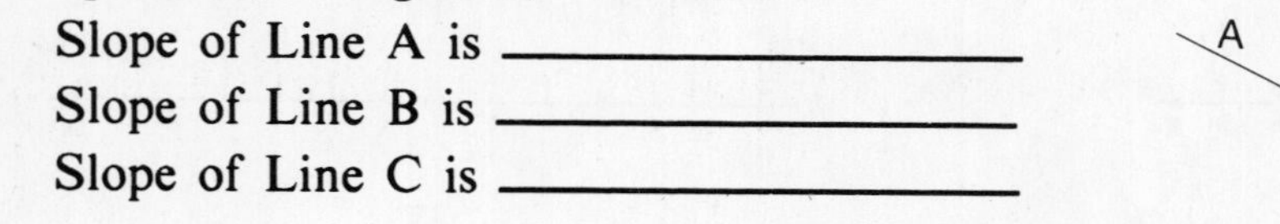

Slope of Line A is ____________________
Slope of Line B is ____________________
Slope of Line C is ____________________

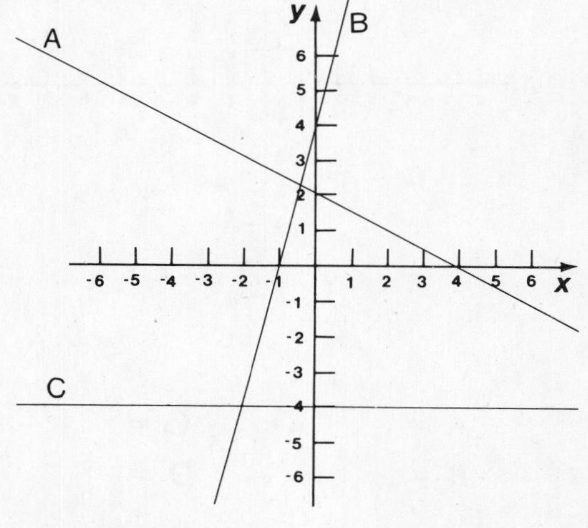

9. Graph the equation $y = -2x + 4$ and identify the x and y intercepts.

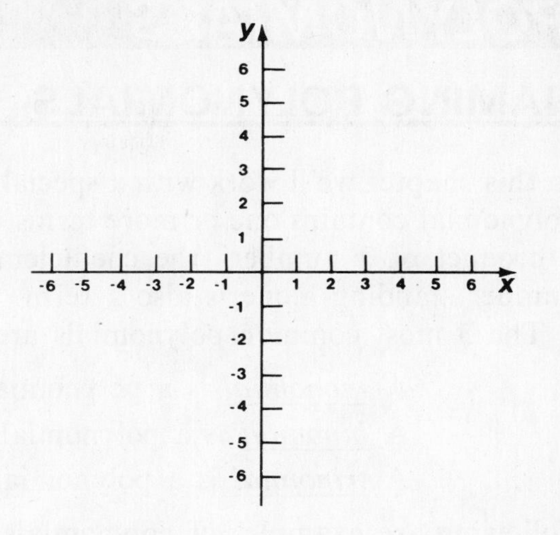

Do your work on the graph at the right.

x intercept =
y intercept =

x	y
0	
+2	
+3	

10. Using the x and y intercepts from problem 9 above, find the slope of the line.

RECTANGULAR COORDINATES SKILLS INVENTORY CHART

Circle the number of any problem that you missed and be sure to review the appropriate practice page. A passing score is 7 correct answers.

Problem Number	Skill Area	Practice Page	Problem Number	Skill Area	Practice Page
1	drawing a graph	84–85	6	plotting points	88
2	ordered pairs	86	7	finding intercepts	90
3	ordered pairs	86	8	slope	92
4	ordered pairs	86	9	graphing equations	94–95
5	ordered pairs	86	10	figuring slope	93

If you missed more than 3 questions, you should review this chapter.

POLYNOMIALS

NAMING POLYNOMIALS

In this chapter we'll work with a special type of algebraic expression called a *polynomial*. A polynomial contains one or more terms combined by addition and subtraction. Each term is a product of a number (the coefficient) times one or more variables with exponents. A number standing alone is also a term.

The 3 most common polynomials are *monomials, binomials*, and *trinomials*.

A *monomial* is a polynomial of one term.
A *binomial* is a polynomial of two terms.
A *trinomial* is a polynomial of three terms.

Following are examples of polynomials:

Monomials	Binomials	Trinomials
z	$x - 3$	$x + y + z$
$-2x$	$3y + 9$	$5r - 3s + 6t$
$3a^2$	$6c^2 + 2c$	$4y^2 - 5y + 7$
$-6x^2y^2$	$4a^2b^2 + 2ab$	$7y^3 - 10y^2 + 9y$

To identify a term, include the sign that precedes it. For example, in the trinomial $3x^2y^2 - 9xy + 8$, the terms are $+3x^2y^2$, $-9xy$, and $+8$.

Name each polynomial as a monomial, binomial, or trinomial.

1. $5x - 6$ ____________ **2.** $4x + 6y - z$ ____________

3. $-y$ ____________ **4.** $-5y^2 + 3y$ ____________

5. $r^2 + s^2 - t^2$ ____________ **6.** $7x^2y^2$ ____________

Identify the terms in the following polynomials.

Examples

$-12x + 7$
$-12x$ and $+7$

$+ 9a - 2b$
$+ 9a$ and $-2b$

$2x^2 - 9x + 4$
$2x^2$, $-9x$, and 4

$- x - y + z$
$-x$, $-y$, and z

7. $-9d + 6$ $3a^2 - 5a$ $-4y - \frac{1}{2}$

8. $-6d + 8e$ $8x^2 + 3y^2$ $\frac{2}{3}z^2 - 7z$

9. $-3y^2 - 7y - 8$ $8x + 4y - 13$

10. $2a + 3b - 2c$ $4u^2 - v^2 + \frac{3}{4}w^2$

RECOGNIZING LIKE TERMS

As you saw on the previous page, a term in a polynomial can be a number standing alone, or it can be the product of a coefficient times one or more variables with exponents. The variables with exponents are called the <u>*variable part*</u> of the term.

EXAMPLE:

$-2x^2y$

coefficient (-2) → variable part (x^2y)

Before studying the arithmetic of polynomials you need to be able to recognize <u>*like terms*</u>. Like terms have identical variable parts. This means that like terms are exactly alike except for their coefficients.

For example, $8x^2y$ and $-2x^2y$ are like terms because the variable part (x^2y) is the same in each term. Terms with different variable parts are called unlike terms.

Here are more examples of like and unlike terms.

Like Terms	**Unlike Terms**
5 and -7	x and y
$-3y$ and $4y$	$-r$ and s
$5x^2$ and $2x^2$	$4x^2$ and $-2z$
$-4a^2b$ and a^2b	a^2b and b^2a

Note: All numbers standing alone are like terms.

For each term below, identify the coefficient (C.) and the variable part (V.P.).

Examples

	C.	V.P.
$-4x$	*−4*	x
$+y^2$	*+1*	y^2
$-9x^2y$	*−9*	x^2y
$3a^2b^2$	*+3*	a^2b^2

		C.	V.P.			C.	V.P.
1.	$+3a$	_____	_____	**2.**	$-5z$	_____	_____
3.	$+z^2$	_____	_____	**4.**	$-a^2$	_____	_____
5.	$-4c^2d$	_____	_____	**6.**	$7uv^2$	_____	_____
7.	$8x^2y^2$	_____	_____	**8.**	$-4ab^3$	_____	_____

Circle the pair of like terms in each group of terms below.

9. $x, x^2, -5x$

10. $a^2, -4a^2, 2a, 5a^4$

11. $+2xy^2, -x^2y, +xy^2, -2x$

12. $c, -2, +4$

13. $x^3, -2x^4, 4x^3, 3x^2$

14. $+3rs, -rs^2, +3r^2, +3rs^2$

15. $y^3, 3y, 2y^3$

16. $+c^2, +2c, -c^3, -5c^2$

17. $x^2y^2, -x^2y, -y^2x, -x^2y^2$

ADDING MONOMIALS

To add monomials, add the coefficients of like terms according to the rules for adding signed numbers and attach (keep) the variables. Only the coefficients of like terms can be added.

EXAMPLE 1. Add $3x$ and $2x$.

$$\begin{array}{r} 3x \\ \underline{2x} \\ \mathbf{5x} \end{array}$$

EXAMPLE 2. Add $-5a^2b$ and a^2b.

$$-5a^2b + a^2b = \mathbf{-4a^2b}$$

Note: The coefficient of a^2b is understood to be $+1$.

Addition of unlike terms is shown by placing the sign of the second term between the two added terms. You cannot add the coefficients of unlike terms.

EXAMPLE 3. Add $4x$ and $7y$.

$$\begin{array}{l} 4x \\ \underline{7y} \\ \mathbf{4x + 7y} \end{array}$$

x and y are unlike terms.

EXAMPLE 4. Add $3c^2d$ and $-2cd$.

$$\begin{array}{r} 3c^2d \\ \underline{-2cd} \\ \mathbf{3c^2d - 2cd} \end{array}$$

c^2d and cd are unlike terms.

Add.

Examples

$$\begin{array}{r} 5x \\ \underline{3x} \\ \mathbf{8x} \end{array} \qquad \begin{array}{r} 7y \\ \underline{-y} \\ \mathbf{6y} \end{array} \qquad \begin{array}{r} -6x \\ \underline{-7x} \\ \mathbf{-13x} \end{array} \qquad \begin{array}{r} 3x^2y \\ \underline{-x^2y} \\ \mathbf{2x^2y} \end{array}$$

1. $\begin{array}{r} 4y \\ \underline{2y} \end{array}$ $\begin{array}{r} 2z \\ \underline{z} \end{array}$ $\begin{array}{r} n \\ \underline{5n} \end{array}$ $\begin{array}{r} 4r \\ \underline{3r} \end{array}$ $\begin{array}{r} 2c \\ \underline{5c} \end{array}$

2. $\begin{array}{r} 9z \\ \underline{-4z} \end{array}$ $\begin{array}{r} -7c \\ \underline{4c} \end{array}$ $\begin{array}{r} -8u \\ \underline{6u} \end{array}$ $\begin{array}{r} 2r \\ \underline{-r} \end{array}$ $\begin{array}{r} 4x \\ \underline{-3x} \end{array}$

3. $\begin{array}{r} -4z \\ \underline{-6z} \end{array}$ $\begin{array}{r} -8y \\ \underline{-5y} \end{array}$ $\begin{array}{r} -6n \\ \underline{-9n} \end{array}$ $\begin{array}{r} -6y \\ \underline{-y} \end{array}$ $\begin{array}{r} -6a \\ \underline{-3a} \end{array}$

4. $\begin{array}{r} -7ab^2 \\ \underline{3ab^2} \end{array}$ $\begin{array}{r} 4cd \\ \underline{-7cd} \end{array}$ $\begin{array}{r} -7xyz \\ \underline{3xyz} \end{array}$ $\begin{array}{r} -5a^2b \\ \underline{-6a^2b} \end{array}$ $\begin{array}{r} 4x^2y^2 \\ \underline{x^2y^2} \end{array}$

Examples

$$\begin{array}{r} -4xy \\ -5xy \\ \underline{3xy} \\ \mathbf{-6xy} \end{array} \qquad \begin{array}{r} x \\ \underline{-y^2} \\ \mathbf{x - y^2} \end{array}$$

5.

$3a^2b$	$-6x^2y$	c^2d	$8xy$	$12z$
$-2a^2b$	$4x^2y$	$3c^2d$	$-xy$	z
a^2b	$7x^2y$	$-5c^2d$	$-2xy$	$7z$

6.

$2c^2$	$-4ab^2$	$2yz$	$7r^2$	$-6t$
$-3d$	$2ab$	xy	$-s^2$	$-4u$

$\mathbf{4x + 6x = 10x}$

$\mathbf{-6a + 3a = -3a}$

$\mathbf{-4a^2 + (-a^2) = -5a^2}$

7. $13yz + 8yz =$ $\qquad$ $9xy^2 + 11xy^2 =$

8. $-5b^2 + 4b^2 =$ $\qquad$ $12c + (-5c) =$

9. $-12xy + (23xy) =$ $\qquad$ $-7d^2 + (-6d^2) =$

More Practice In Adding Monomials.

10.

$-8x$	$4x^2$	$-7y$	$3r^2$	$-5xy$	uv
$-2x$	$-5x^2$	$3z$	r^2	$-2xy$	$4uv$

11.

$7c$	$-6b^2$	$4a^2b$	$-u^2$	x^2y	$-8v$
$-5d$	$-2b^2$	$3a^2b$	$4u^2$	$-6x^2y$	$3v$

12.

$5xy^2$	$4b$	$2ab$	$12a$	$-8v$	$-3r$
$-2xy^2$	$-b$	$3ad$	$8a$	$-5v$	$12r$

13.

$7a^2$	$9xy$	$-6cd$	$7r^2$	$-6xy^2$	$5ab$
$5a^2$	$-6xy$	$-4cd$	$4r^2$	$-8xy^2$	$3ab$
$3a^2$	$-xy$	$-7cd$	$-5r^2$	$3xy^2$	$-9ab$

14. $8x^2 + (-5x^2) =$ $\qquad$ $-7ab + 5ab =$ $\qquad$ $12xyz + 6xyz =$

15. $-9a^2b + (-6a^2b) =$ $\qquad$ $5x + 4x =$ $\qquad$ $9yz + (-yz) =$

16. $a^2b + (-a^2b) =$ $\qquad$ $13x^2y^2 + x^2y^2 =$ $\qquad$ $-6y + (-2y) =$

ADDING POLYNOMIALS

To add polynomials, add the coefficients of like terms according to the rules for adding signed numbers and attach the variables.

EXAMPLE 1: Add a monomial and a binomial.

Add:
$$\begin{array}{l} 3x - 4 \\ \underline{2x \phantom{{}-4}} \\ \mathbf{5x - 4} \end{array}$$

EXAMPLE 2: Add two binomials with one variable.

Add:
$$\begin{array}{l} 4y + 5 \\ \underline{2y - 8} \\ \mathbf{6y - 3} \end{array}$$

EXAMPLE 3: Add two binomials with two variables.

Add:
$$\begin{array}{l} 2a - b \\ \underline{a + 2b} \\ \mathbf{3a + b} \end{array}$$

EXAMPLE 4: Add two trinomials.

Add:
$$\begin{array}{l} 3x^2 - 7x + 8 \\ \underline{2x^2 + 5x - 6} \\ \mathbf{5x^2 - 2x + 2} \end{array}$$

Parentheses are often used in addition and subtraction to separate polynomials. When there is a plus sign or no sign at all in front of the parentheses, just remove the parentheses. Do not change the signs of the terms that were inside the parentheses.

EXAMPLE 5. Add: $(2x - 4y) + (-3x + 5y)$.

Add: $(2x - 4y) + (-3x + 5y)$

Step 1. Remove the parentheses from both binomials. Leave the sign of each term unchanged.

$2x - 4y \; + \; -3x + 5y$

Step 2. Combine like terms.
$2x - 3x = -x$
$-4y + 5y = +y$

x terms

$= 2x - 4y - 3x + 5y$

y terms

Answer: $-x + y$

$= -x + y$

Add.

Examples

$$\begin{array}{l} 2x + 9 \\ \underline{ - 4} \\ 2x + 5 \end{array}$$

$$\begin{array}{l} 5y + 6 \\ \underline{3y \phantom{{}+6}} \\ 8y + 6 \end{array}$$

1. $\begin{array}{l} 3y + 5 \\ \underline{ - 9} \end{array}$ $\quad$ $\begin{array}{l} 5z - 7 \\ \underline{ + 5} \end{array}$ $\quad$ $\begin{array}{l} 3a + 1 \\ \underline{ + 4} \end{array}$ $\quad$ $\begin{array}{l} 7u - 2 \\ \underline{ - 7} \end{array}$

2. $\begin{array}{l} 3x - 5 \\ \underline{x \phantom{{}-5}} \end{array}$ $\quad$ $\begin{array}{l} 4a + 8 \\ \underline{2a \phantom{{}+8}} \end{array}$ $\quad$ $\begin{array}{l} -6b + 3 \\ \underline{7b \phantom{{}+3}} \end{array}$ $\quad$ $\begin{array}{l} 4x - 7 \\ \underline{-6x \phantom{{}-7}} \end{array}$

Examples

$$\begin{array}{r} 4x + 5 \\ \underline{x + 3} \\ 5x + 8 \end{array}$$

3. $\begin{array}{r} 3y + 2 \\ \underline{y + 4} \end{array}$ $\quad \begin{array}{r} z + 1 \\ \underline{z + 9} \end{array}$ $\quad \begin{array}{r} 4c + 6 \\ \underline{3c + 1} \end{array}$ $\quad \begin{array}{r} 5x + 3 \\ \underline{2x + 8} \end{array}$

$$\begin{array}{r} 5y + 2 \\ \underline{3y - 4} \\ 8y - 2 \end{array}$$

4. $\begin{array}{r} 2z - 6 \\ \underline{z + 3} \end{array}$ $\quad \begin{array}{r} 7c + 2 \\ \underline{3c - 1} \end{array}$ $\quad \begin{array}{r} 3a + 4 \\ \underline{2a - 4} \end{array}$ $\quad \begin{array}{r} 7u - 8 \\ \underline{-3u + 6} \end{array}$

$$\begin{array}{r} 3x^2 + 5x \\ \underline{2x^2 - 4x} \\ 5x^2 + x \end{array}$$

5. $\begin{array}{r} 6y^2 - 5y \\ \underline{3y^2 + 2y} \end{array}$ $\quad \begin{array}{r} 2a^2 + 4a \\ \underline{a^2 + 2a} \end{array}$ $\quad \begin{array}{r} 6u^2 + 4v \\ \underline{-4u^2 + 2v} \end{array}$ $\quad \begin{array}{r} 4r^2 - 6r \\ \underline{r^2 + 6r} \end{array}$

$$\begin{array}{r} 4x + 3y - 2z \\ \underline{3x - 2y + 5z} \\ 7x + y + 3z \end{array}$$

6. $\begin{array}{r} 2a - 3b + 7c \\ \underline{3a + b - 3c} \end{array}$ $\quad \begin{array}{r} -4r + 2s - 6t \\ \underline{5r + 3s + 5t} \end{array}$ $\quad \begin{array}{r} 5x + 2y + 8z \\ \underline{3x - 4y - 7z} \end{array}$

$(4x + 5y) + (-3x + 7y) = 4x + 5y - 3x + 7y = \mathbf{x + 12y}$

7. $(3a + 6b) + (-5a + 2b) =$

$(2r - 3s) + (4r + 3s) =$

$(7x + 4y) + (3x - 2y) =$

$(3a - 5b) + (2a - 5b) =$

$(2x + 3y + z) + (x - y + z) = 2x + 3y + z + x - y + z = \mathbf{3x + 2y + 2z}$

8. $(4x + 2y + z) + (3x - y + 2z) =$

$(3x - y - 2z) + (5x + 2y + z) =$

$(2x^2 - x + 3) + (-x^2 + 3x - 4) =$

$(y^2 + 3y - 5) + (2y^2 - y + 7) =$

SUBTRACTING MONOMIALS

To subtract monomials, subtract the coefficients of like terms according to the rules for signed numbers. As in addition, only the coefficients of like terms can be subtracted.

EXAMPLE 1. Subtract $9y$ from $5y$.

$$\begin{array}{lll} & & \textit{add} \\ & 5y & \left[\begin{array}{r} 5y \\ \underline{-9y} \\ \mathbf{-4y} \end{array}\right] \\ \textbf{Subtract:} & \underline{9y} & \end{array}$$

Remember to change the sign of the number being subtracted and then follow the rules for adding signed numbers.

EXAMPLE 2. Subtract $-3x^2$ from $-x^2$.

Subtract: $-x^2 - (-3x^2)$

$= -x^2 + 3x^2$

$= \mathbf{2x^2}$

Note: The coefficient of $-x^2$ is understood to be -1.

To show subtraction of unlike terms, change the sign of the term being subtracted and follow the rule for addition of unlike terms.

EXAMPLE 3. Subtract $4c$ from $7d$.

$$\begin{array}{lll} & & \textit{add} \\ & 7d & \left[\begin{array}{r} 7d \\ \underline{-4c} \\ \mathbf{7d - 4c} \end{array}\right] \\ \textbf{Subtract:} & \underline{4c} & \end{array}$$

c and d are unlike variables.

EXAMPLE 4. Subtract $-6a$ from $-8b$.

$$\begin{array}{lll} & & \textit{add} \\ & -8b & \left[\begin{array}{r} 8b \\ \underline{6a} \\ \mathbf{8b + 6a} \end{array}\right] \\ \textbf{Subtract:} & \underline{-6a} & \end{array}$$

a and b are unlike variables.

Subtract.

Examples

$$\begin{array}{rl} & \textit{add} \\ 7y & \left[\begin{array}{r} 7y \\ \underline{-y} \\ 6y \end{array}\right] \\ \underline{y} & \end{array} \qquad \begin{array}{rl} & \textit{add} \\ 8x & \left[\begin{array}{r} 8x \\ \underline{3x} \\ 11x \end{array}\right] \\ \underline{-3x} & \end{array} \qquad \begin{array}{rl} & \textit{add} \\ -6z & \left[\begin{array}{r} -6z \\ \underline{7z} \\ z \end{array}\right] \\ \underline{-7z} & \end{array}$$

1.	$\begin{array}{r} 5z \\ \underline{8z} \end{array}$	$\begin{array}{r} 9a \\ \underline{8a} \end{array}$	$\begin{array}{r} 12c \\ \underline{7c} \end{array}$	$\begin{array}{r} 3a \\ \underline{6a} \end{array}$	$\begin{array}{r} x \\ \underline{5x} \end{array}$
2.	$\begin{array}{r} 9a \\ \underline{-5a} \end{array}$	$\begin{array}{r} -3n \\ \underline{7n} \end{array}$	$\begin{array}{r} 11r \\ \underline{-5r} \end{array}$	$\begin{array}{r} -7z \\ \underline{8z} \end{array}$	$\begin{array}{r} -4y \\ \underline{9y} \end{array}$
3.	$\begin{array}{r} -8c \\ \underline{-9c} \end{array}$	$\begin{array}{r} -11y \\ \underline{-6y} \end{array}$	$\begin{array}{r} -d \\ \underline{-3d} \end{array}$	$\begin{array}{r} -4s \\ \underline{-s} \end{array}$	$\begin{array}{r} -12x \\ \underline{-12x} \end{array}$

Examples

$\begin{array}{r} 5x^2y \\ \underline{-2x^2y} \end{array} \quad \overset{add}{\left[\begin{array}{r} 5x^2y \\ \underline{2x^2y} \\ 7x^2y \end{array}\right]}$

$\begin{array}{l} a \\ \underline{b} \end{array} \quad \overset{add}{\left[\begin{array}{r} a \\ \underline{-b} \\ a-b \end{array}\right]}$

$6x - 5x = x$

4. $\begin{array}{r} 3a^2b^2 \\ \underline{-a^2b^2} \end{array}$ $\quad \begin{array}{r} -8xyz \\ \underline{5xyz} \end{array}$ $\quad \begin{array}{r} -7n^2 \\ \underline{-8n^2} \end{array}$ $\quad \begin{array}{r} -11x^2 \\ \underline{-11x^2} \end{array}$ $\quad \begin{array}{r} 5x^2y \\ \underline{3x^2y} \end{array}$

5. $\begin{array}{r} 3x^2 \\ \underline{x^3} \end{array}$ $\quad \begin{array}{r} -5cd \\ \underline{8c^2} \end{array}$ $\quad \begin{array}{r} 6r^2s \\ \underline{-2rs} \end{array}$ $\quad \begin{array}{r} -9u^2v \\ \underline{uv^2} \end{array}$ $\quad \begin{array}{r} -12x^2 \\ \underline{-8x} \end{array}$

6. $9y - 3y =$ $\qquad 7a^2b - 4a^2b =$

$8a - (-3a) = 11a$

$-5y - (-7y) = 2y$

$10x - 5x - 3x = 2x$

$15y - (-4y) - 3y = 16y$

7. $9b - (-4b) =$ $\qquad 12x - (-7x) =$

8. $-8x - (-3x) =$ $\qquad -9z - (-12z) =$

9. $12y - 6y - 4y =$ $\qquad 14z - 7z - 3z =$

10. $20x - (-6x) - 9x =$ $\qquad 8z - 6z - (-z) =$

More Practice In Subtracting Monomials

11. $\begin{array}{r} 8z \\ \underline{5z} \end{array}$ $\quad \begin{array}{r} 5x^2 \\ \underline{7xy} \end{array}$ $\quad \begin{array}{r} -9c^2d \\ \underline{8c^2d} \end{array}$ $\quad \begin{array}{r} 4xy \\ \underline{-6xy} \end{array}$ $\quad \begin{array}{r} 8uv \\ \underline{5uv} \end{array}$ $\quad \begin{array}{r} -7y \\ \underline{-8y} \end{array}$

12. $\begin{array}{r} x^2y^2 \\ \underline{-x^2y} \end{array}$ $\quad \begin{array}{r} 6y \\ \underline{-9y} \end{array}$ $\quad \begin{array}{r} 9c \\ \underline{6c} \end{array}$ $\quad \begin{array}{r} 12r^2 \\ \underline{-9r^2} \end{array}$ $\quad \begin{array}{r} -7y^2 \\ \underline{-8y^2} \end{array}$ $\quad \begin{array}{r} 4x \\ \underline{5x} \end{array}$

13. $\begin{array}{r} -8y \\ \underline{-7y} \end{array}$ $\quad \begin{array}{r} 4x \\ \underline{18x} \end{array}$ $\quad \begin{array}{r} 13z \\ \underline{9z} \end{array}$ $\quad \begin{array}{r} 8x^2 \\ \underline{-9x^2} \end{array}$ $\quad \begin{array}{r} -5y^2 \\ \underline{-8y^2} \end{array}$ $\quad \begin{array}{r} -11z \\ \underline{23z} \end{array}$

14. $\begin{array}{r} 13x^3 \\ \underline{-9x^2} \end{array}$ $\quad \begin{array}{r} 5c^2d \\ \underline{4c^2d} \end{array}$ $\quad \begin{array}{r} 8rs \\ \underline{-rs^2} \end{array}$ $\quad \begin{array}{r} -9t^2 \\ \underline{8t^2} \end{array}$ $\quad \begin{array}{r} -14x \\ \underline{-5x} \end{array}$ $\quad \begin{array}{r} 14u \\ \underline{7u} \end{array}$

15. $-9x - (-6x) =$ $\qquad 12y - 7y =$ $\qquad 26x^2 - (-4x^2) =$

16. $12x - 9x - 4x =$ $\qquad 5y^2 - y^2 - 3y^2 =$ $\qquad 14t - 6t =$

SUBTRACTING POLYNOMIALS

To subtract polynomials, change the signs of all the terms being subtracted. Then follow the rules for adding signed numbers.

EXAMPLE 1. Subtract $3x - 4$ from $5x - 6$.

Step 1. Change the signs of $3x - 4$.
$3x - 4$ becomes $-3x + 4$.

Step 2. Add $5x - 6$ and $-3x + 4$.

Answer: $\mathbf{2x - 2}$

$$\begin{array}{r} 5x - 6 \\ \underline{3x - 4} \end{array} = \overset{\textit{add}}{\left[\begin{array}{r} 5x - 6 \\ \underline{-3x + 4} \\ \mathbf{2x - 2} \end{array}\right]}$$

When a subtraction sign precedes a polynomial in parentheses, remove the parentheses and change the signs of all the terms being subtracted.

EXAMPLE 2. Subtract: $(3a - 4b) - (2a - 5b)$

Step. 1. Remove the parentheses from both binomials. Change the sign of each term in the binomial being subtracted.

Step. 2. Combine like terms:
$3a - 2a = +a$
$-4b + 5b = +b$

Answer: $\boldsymbol{a + b}$

Subtract: $(3a - 4b) - (2a - 5b)$
$= 3a - 4b - 2a + 5b$

a terms
$= 3a - 4b - 2a + 5b$
b terms
$= \boldsymbol{a + b}$

Subtract.

Examples

$$\begin{array}{r} 3y - 8 \\ \underline{- 4} \end{array} \quad \overset{\textit{add}}{\left[\begin{array}{r} 3y - 8 \\ \underline{+ 4} \\ 3y - 4 \end{array}\right]}$$

$$\begin{array}{r} 4x + 9 \\ \underline{-3x \quad} \end{array} \quad \overset{\textit{add}}{\left[\begin{array}{r} 4x + 9 \\ \underline{3x \quad} \\ 7x + 9 \end{array}\right]}$$

$$\begin{array}{r} 5a + 2 \\ \underline{a + 2} \end{array} \quad \overset{\textit{add}}{\left[\begin{array}{r} 5a + 2 \\ \underline{-a - 2} \\ \mathbf{4a} \quad \end{array}\right]}$$

1. $\begin{array}{r} 2z + 6 \\ \underline{7} \end{array}$ $\qquad \begin{array}{r} 7n + 3 \\ \underline{6} \end{array}$ $\qquad \begin{array}{r} 2s - 5 \\ \underline{- 7} \end{array}$ $\qquad \begin{array}{r} 4x + 5 \\ \underline{8} \end{array}$

2. $\begin{array}{r} 7y - 5 \\ \underline{2y \quad} \end{array}$ $\qquad \begin{array}{r} 3x + 7 \\ \underline{-4x \quad} \end{array}$ $\qquad \begin{array}{r} 6y - 8 \\ \underline{-3y \quad} \end{array}$ $\qquad \begin{array}{r} 3z + 2 \\ \underline{2z \quad} \end{array}$

3. $\begin{array}{r} 3x - 6 \\ \underline{x - 4} \end{array}$ $\qquad \begin{array}{r} 2y - 6 \\ \underline{-y + 7} \end{array}$ $\qquad \begin{array}{r} 4z + 9 \\ \underline{3z + 4} \end{array}$ $\qquad \begin{array}{r} 12x + 1 \\ \underline{5x - 7} \end{array}$

$$\begin{array}{r} 4x - 5y \\ \underline{2x + 6y} \end{array} \quad \overset{\textit{add}}{\left[\begin{array}{r} 4x - 5y \\ \underline{-2x - 6y} \\ 2x - 11y \end{array}\right]}$$

4. $\begin{array}{r} 5a - 3b \\ \underline{2a - 7b} \end{array}$ $\quad$ $\begin{array}{r} 12r + 3s \\ \underline{5r - 5s} \end{array}$ $\quad$ $\begin{array}{r} 14y + 4z \\ \underline{7y - 3z} \end{array}$ $\quad$ $\begin{array}{r} 9u - 5 \\ \underline{-u + 3} \end{array}$

$$\begin{array}{r} 2a^2 + 3a \\ \underline{a^2 + 4a} \end{array} \quad \overset{\textit{add}}{\left[\begin{array}{r} 2a^2 + 3a \\ \underline{-a^2 - 4a} \\ a^2 - a \end{array}\right]}$$

5. $\begin{array}{r} 4y^2 - 2y \\ \underline{3y^2 + 4y} \end{array}$ $\quad$ $\begin{array}{r} 6r^2 - 4r \\ \underline{3r^2 - 5r} \end{array}$ $\quad$ $\begin{array}{r} 9u^2 + 3u \\ \underline{2u^2 - 4u} \end{array}$ $\quad$ $\begin{array}{r} 6x^2 + 2 \\ \underline{x^2} \end{array}$

$$\begin{array}{r} 4x^2 - 5x + 7 \\ \underline{3x^2 + 6x - 7} \end{array} \quad \overset{\textit{add}}{\left[\begin{array}{r} 4x^2 - 5x + 7 \\ \underline{-3x^2 - 6x + 7} \\ x^2 - 11x + 14 \end{array}\right]}$$

6. $\begin{array}{r} 3a^2 - 4a + 1 \\ \underline{a^2 + 3a + 5} \end{array}$ $\quad$ $\begin{array}{r} 8r^2 + 7r - 4 \\ \underline{3r^2 + 6r - 5} \end{array}$ $\quad$ $\begin{array}{r} 3x^2 + 2x \\ \underline{2x^2 - 6x + 8} \end{array}$

$$(5x + 3y) - (2x - 5y) = 5x + 3y - 2x + 5y = \mathbf{3x + 8y}$$

7. $(7a - 4b) - (2a + 3b) =$

$(5r - 3s) - (-3r - 2s) =$

$(2u + v) - (u + v) =$

$(7x - 3y) - (x + y) =$

$$(2x + 5y - 2z) - (4x - 3y + 2z) = 2x + 5y - 2z - 4x + 3y - 2z = \mathbf{-2x + 8y - 4z}$$

8. $(5x + 3y - 2z) - (3x + 2y - 6z) =$

$(x - y - z) - (x + y - z) =$

$(3x^2 - x + 4) - (2x^2 + 3x - 5) =$

$(-a^2 + 4a - 2) - (3a^2 - 5a + 6) =$

$$(-3x + 2y) - (-4x - 3y) = -3x + 2y + 4x + 3y = \mathbf{x + 5y}$$

9. $(6r + 2s) - (-4r + 2s) =$

$(5x - 3y + 7z) - (2x - y - z) =$

$(7x^2 - 2x + 4) - (3x^2 - 2x - 4) =$

$(4a - 2b + 3c) - (a - b - 2c) =$

MULTIPLYING A MONOMIAL BY A NUMBER

To multiply a monomial by a number, multiply the coefficient times the number and attach the variables. Follow the rules for multiplying signed numbers.

EXAMPLE 1. Multiply $3x$ by 4.

$$\begin{array}{r} 3x \\ \underline{4} \\ \mathbf{12x} \end{array}$$

EXAMPLE 2. Multiply: $-2(4a^2b)$

$-2(4a^2b) = \mathbf{-8a^2b}$

Multiply.

Examples

$$\begin{array}{r} 4x^2 \\ \underline{3} \\ 12x^2 \end{array} \qquad \begin{array}{r} -c^2d \\ \underline{8} \\ -8c^2d \end{array}$$

1. $\begin{array}{r} 4y \\ \underline{2} \end{array}$ $\quad \begin{array}{r} 7z \\ \underline{-3} \end{array}$ $\quad \begin{array}{r} -3a \\ \underline{6} \end{array}$ $\quad \begin{array}{r} 5c \\ \underline{-7} \end{array}$ $\quad \begin{array}{r} -4b \\ \underline{8} \end{array}$

2. $\begin{array}{r} 7xy \\ \underline{-2} \end{array}$ $\quad \begin{array}{r} -2ab^2 \\ \underline{6} \end{array}$ $\quad \begin{array}{r} 8rs \\ \underline{4} \end{array}$ $\quad \begin{array}{r} 9y^2 \\ \underline{3} \end{array}$ $\quad \begin{array}{r} 12z^3 \\ \underline{-5} \end{array}$

$(2x^2)(-3) = -6x^2$

$(4a^2b^2)(-\frac{1}{2}) = -2a^2b^2$

3. $(-4x^2y)(+4) =$ $\qquad (-6a^2b^2)(-5) =$

4. $(6xyz)(-\frac{1}{3}) =$ $\qquad (12y^2z)(\frac{1}{4}) =$

Multiply.

5. $\begin{array}{r} 5a \\ \underline{-3} \end{array}$ $\quad \begin{array}{r} -9x^2 \\ \underline{5} \end{array}$ $\quad \begin{array}{r} 5x^2y^3 \\ \underline{-7} \end{array}$ $\quad \begin{array}{r} 6x \\ \underline{3} \end{array}$ $\quad \begin{array}{r} 3xy^2 \\ \underline{4} \end{array}$ $\quad \begin{array}{r} 12a \\ \underline{-2} \end{array}$

6. $\begin{array}{r} 12rs \\ \underline{-4} \end{array}$ $\quad \begin{array}{r} 9abc \\ \underline{3} \end{array}$ $\quad \begin{array}{r} 6r^2 \\ \underline{-5} \end{array}$ $\quad \begin{array}{r} 6n \\ \underline{3} \end{array}$ $\quad \begin{array}{r} 5r^2s^2 \\ \underline{7} \end{array}$ $\quad \begin{array}{r} -8b \\ \underline{-6} \end{array}$

7. $(4xy)(-\frac{1}{4}) =$ $\qquad (5xy^2)(-4) =$ $\qquad (12x^2y^2)(-\frac{5}{3}) =$

8. $(14ab^2)(3) =$ $\qquad (25c^2d)(-\frac{1}{5}) =$ $\qquad (13xyz)(4) =$

EXPONENTS IN MULTIPLICATION

Multiplication often involves terms with *like variables* (same letters). As the following examples show, a shortcut to multiplying like variables is to add exponents.

EXAMPLE 1. Multiply x^4 by x^3.

The *long way* is to write out each term and then count x's.

$$x^4 \cdot x^3 = \underbrace{x \cdot x \cdot x \cdot x \cdot}_{x^4}\underbrace{x \cdot x \cdot x \cdot}_{x^3} = x^7$$

since there are 7 x's in a row being multiplied.

Answer: x^7

A *shortcut* is to add the exponents of each term as follows:

$$x^4 \cdot x^3 = x^{4+3} = x^7$$

Seeing this, you can now write the rule for multiplying like variables.

RULE: To multiply like variables, add the exponents of that variable.

EXAMPLE 2. Multiply y^3 by y^2.

$$\begin{array}{l} y^3 \\ \underline{y^2} \\ y^{3+2} = y^5 \end{array}$$

EXAMPLE 3. Multiply: $(z^2)(z)$

$(z^2)(z) = z^{2+1} = z^3$

Remember: *If no exponent is written, it is understood to be 1.*

Use the method of adding exponents to multiply the following terms.

Examples

$$\begin{array}{l} x \\ \underline{x} \\ x^2 \end{array} \qquad \begin{array}{l} s^2 \\ \underline{s} \\ s^3 \end{array}$$

$x(x) = x^2$

$(y^2)(y) = y^3$

1. $\begin{array}{l} y \\ \underline{y} \end{array}$ $\qquad \begin{array}{l} z \\ \underline{z} \end{array}$ $\qquad \begin{array}{l} a \\ \underline{a} \end{array}$ $\qquad \begin{array}{l} c \\ \underline{c} \end{array}$ $\qquad \begin{array}{l} d \\ \underline{d} \end{array}$

2. $\begin{array}{l} x^3 \\ \underline{x} \end{array}$ $\qquad \begin{array}{l} a^2 \\ \underline{a} \end{array}$ $\qquad \begin{array}{l} z^3 \\ \underline{z} \end{array}$ $\qquad \begin{array}{l} d^4 \\ \underline{d} \end{array}$ $\qquad \begin{array}{l} c^2 \\ \underline{c} \end{array}$

3. $(y)(y) =$ $\qquad z(z) =$ $\qquad (a)(a) =$

4. $(x^3)(x) =$ $\qquad a^2(a) =$ $\qquad b^3(b) =$

MULTIPLYING MONOMIALS

To multiply monomials, follow these two steps:

Step 1. Multiply the coefficients according to the rules for multiplying signed numbers.

Step 2. Multiply the variables by adding the exponents of each like variable. Write the variables in alphabetical order.

EXAMPLE 1. Multiply $3x^2$ by $-4x$.

Step 1. Multiply the coefficients: $(3)(-4) = -12$

Step 2. Add the exponents of the x variable.

$x^2 \cdot x = x^{2+1} = x^3$

Answer: $\mathbf{-12x^3}$

$$\begin{array}{r} 3x^2 \\ \underline{-\ 4x} \\ \mathbf{-12x^3} \end{array}$$

EXAMPLE 2. Multiply: $-2a^2b^2(-3ab^3)$

$-2a^2b^2(-3ab^3) = (-2)(-3)a^{2+1}b^{2+3} = \mathbf{6a^3b^5}$

Step 1. Multiply the coefficients. $(-2)(-3) = 6$

Step 2. Add the exponents of each like variable.

$a^2a = a^{2+1} = a^3$ and $b^2b^3 = b^{2+3} = \mathbf{b^5}$

Answer: $\mathbf{6a^3b^5}$

Multiply.

Examples

$$\begin{array}{r} 4x \\ \underline{2x} \\ 8x^2 \end{array} \qquad \begin{array}{r} 5c \\ \underline{-2d} \\ -10cd \end{array} \qquad \begin{array}{r} -3x^2 \\ \underline{2y} \\ -6x^2y \end{array} \qquad \begin{array}{r} 2xy \\ \underline{3x} \\ 6x^2y \end{array}$$

1.	$\begin{array}{r} 3y \\ \underline{2y} \end{array}$	$\begin{array}{r} -2z \\ \underline{z} \end{array}$	$\begin{array}{r} 5y \\ \underline{-2y} \end{array}$	$\begin{array}{r} 7c \\ \underline{-4c} \end{array}$	$\begin{array}{r} -6x \\ \underline{3x} \end{array}$
2.	$\begin{array}{r} 2x \\ \underline{3y} \end{array}$	$\begin{array}{r} 6a \\ \underline{3b} \end{array}$	$\begin{array}{r} -2n \\ \underline{4m} \end{array}$	$\begin{array}{r} -4z \\ \underline{-2x} \end{array}$	$\begin{array}{r} -4b \\ \underline{2r} \end{array}$
3.	$\begin{array}{r} 2a^2 \\ \underline{3b} \end{array}$	$\begin{array}{r} -3c^2 \\ \underline{-2d} \end{array}$	$\begin{array}{r} 7x^2 \\ \underline{-4z} \end{array}$	$\begin{array}{r} -4m^2 \\ \underline{-2n} \end{array}$	$\begin{array}{r} 3z^2 \\ \underline{y} \end{array}$
4.	$\begin{array}{r} -4ac \\ \underline{2c} \end{array}$	$\begin{array}{r} 5yz \\ \underline{-8y} \end{array}$	$\begin{array}{r} 8rs \\ \underline{2r} \end{array}$	$\begin{array}{r} 3xy \\ \underline{-2x} \end{array}$	$\begin{array}{r} -6yz \\ \underline{-3y} \end{array}$

Examples

$\begin{array}{r}4x^4y\\ \underline{-2xz}\\ -8x^5yz\end{array}$

$\begin{array}{r}-3x^2y^2\\ \underline{-2xy}\\ 6x^3y^3\end{array}$

5.	$\begin{array}{r}3y^2z\\ \underline{4xy}\end{array}$	$\begin{array}{r}-6x^2z\\ \underline{-2xy}\end{array}$	$\begin{array}{r}-2ab^2\\ \underline{ac}\end{array}$	$\begin{array}{r}-5rs^2\\ \underline{2rt}\end{array}$	$\begin{array}{r}8a^2b\\ \underline{-2ad}\end{array}$
6.	$\begin{array}{r}4ab^2\\ \underline{3a^2b}\end{array}$	$\begin{array}{r}-2r^2st\\ \underline{-4rst^2}\end{array}$	$\begin{array}{r}9r^3st^2\\ \underline{2r^2st}\end{array}$	$\begin{array}{r}8xyz^3\\ \underline{3xy^2}\end{array}$	$\begin{array}{r}-\ a^3b^3\\ \underline{2ab^2}\end{array}$

$(3x^2y)(-2xy) = -6x^3y^2$

$(-x^2y^3)(5xyz) = -5x^3y^4z$

7. $(-4ab^2)(3ab) =$ $\qquad$ $(5r^2s)(-2rs) =$

8. $(3r^3s^2)(2r^2t) =$ $\qquad$ $(2xz^3)(-4y^2z) =$

Multiply the following monomials.

9.	$\begin{array}{r}3a\\ \underline{2a}\end{array}$	$\begin{array}{r}5a\\ \underline{3b}\end{array}$	$\begin{array}{r}-3x^2\\ \underline{4y}\end{array}$	$\begin{array}{r}5r^2s\\ \underline{-3r}\end{array}$	$\begin{array}{r}-7x\\ \underline{-3x}\end{array}$	$\begin{array}{r}4a^2b^2\\ \underline{ab}\end{array}$
10.	$\begin{array}{r}-6a^2\\ \underline{-2b}\end{array}$	$\begin{array}{r}14x^2y\\ \underline{3y}\end{array}$	$\begin{array}{r}5a^3b\\ \underline{4ab^2}\end{array}$	$\begin{array}{r}12y\\ \underline{2y}\end{array}$	$\begin{array}{r}13rs^2\\ \underline{-\ 3t}\end{array}$	$\begin{array}{r}-6x\\ \underline{-3y}\end{array}$
11.	$\begin{array}{r}-6y\\ \underline{2z}\end{array}$	$\begin{array}{r}5a^2b\\ \underline{-3ac}\end{array}$	$\begin{array}{r}7z\\ \underline{-2z}\end{array}$	$\begin{array}{r}9ab^3\\ \underline{4a}\end{array}$	$\begin{array}{r}4r^2s^2\\ \underline{4rs}\end{array}$	$\begin{array}{r}-4b\\ \underline{3b}\end{array}$
12.	$\begin{array}{r}7x^2y^2\\ \underline{2x}\end{array}$	$\begin{array}{r}7a\\ \underline{4b}\end{array}$	$\begin{array}{r}-6uv^3\\ \underline{2u^2v}\end{array}$	$\begin{array}{r}9x\\ \underline{3z}\end{array}$	$\begin{array}{r}4x^2y\\ \underline{-2xz}\end{array}$	$\begin{array}{r}8a^2b\\ \underline{3c}\end{array}$

13. $(2xy)(-3x^2) =$ $\qquad$ $(-4ab^3)(-a^2b^2) =$ $\qquad$ $(3xyz^2)(2xyz) =$

14. $(-3abc)(2a^2b^2) =$ $\qquad$ $(-4x^2y^2)(2xy) =$ $\qquad$ $(6x^2yz)(-2xy) =$

MULTIPLYING A POLYNOMIAL BY A MONOMIAL

To multiply a binomial or trinomial by a monomial, multiply each term and combine the separate terms.

EXAMPLE 1. Multiply $2x^2 + 4$ by $3x$.

Step 1. Multiply each term of the binomial $2x^2 + 4$ by $3x$.
$3x \cdot 2x^2 = 6x^3$ and $3x \cdot 4 = 12x$

$$\begin{array}{r} 2x^2 + 4 \\ \underline{\quad 3x} \\ 6x^3 + 12x \end{array}$$

Step 2. Combine the separate terms. $6x^3 + 12x$

Answer: $\mathbf{6x^3 + 12x}$

As you have seen, parentheses are often used in algebra to indicate multiplication. Remove parentheses by multiplying each term within the parentheses by the monomial.

EXAMPLE 2. Multiply: $2a\ (4a^2 - 3a + 5)$

$$2a\ (4a^2 - 3a + 5) = 2a \cdot 4a^2 + 2a(-3a) + 2a \cdot 5$$
$$= 8a^3 - 6a^2 + 10a$$

Step 1. Multiply each term of the trinomial $4a^2 - 3a + 5$ by $2a$.
$2a \cdot 4a^2 = 8a^3$ and $2a(-3a) = -6a^2$ and $2a \cdot 5 = 10a$

Step 2. Combine the separate terms. $8a^3 - 6a^2 + 10a$

Answer: $\mathbf{8a^3 - 6a^2 + 10a}$

Multiply.

Examples

$$\begin{array}{r} 7x + 4 \\ \underline{3} \\ 21x + 12 \end{array} \qquad \begin{array}{r} 5y^2 + 9y \\ \underline{-2} \\ -10y^2 - 18y \end{array} \qquad \begin{array}{r} 5x^2 - 3x \\ \underline{2x} \\ 10x^3 - 6x^2 \end{array}$$

1. $\begin{array}{r} 6a + 3 \\ \underline{5} \end{array}$ $\quad \begin{array}{r} 3y - 6 \\ \underline{7} \end{array}$ $\quad \begin{array}{r} 8z + 5 \\ \underline{-4} \end{array}$ $\quad \begin{array}{r} 12x - 4 \\ \underline{8} \end{array}$

2. $\begin{array}{r} 4x^2 - 2x \\ \underline{3} \end{array}$ $\quad \begin{array}{r} 7r^2 + 2r \\ \underline{8} \end{array}$ $\quad \begin{array}{r} 4z^2 - z \\ \underline{-7} \end{array}$ $\quad \begin{array}{r} 5a^2 - 2a \\ \underline{4} \end{array}$

3. $\begin{array}{r} 4a^2 + 4a \\ \underline{-5a} \end{array}$ $\quad \begin{array}{r} 13z^2 - z \\ \underline{-2z} \end{array}$ $\quad \begin{array}{r} 7b^2 + 4b \\ \underline{2b} \end{array}$ $\quad \begin{array}{r} 8r^2 + 6r \\ \underline{-3r} \end{array}$

Examples

$$\begin{array}{r} 4x^2 - 3x + 7 \\ 4 \\ \hline 16x^2 - 12x + 28 \end{array}$$

$$\begin{array}{r} -5a^2 + 6a - 4 \\ 3a \\ \hline -15a^3 + 18a^2 - 12a \end{array}$$

4. $\begin{array}{r} 5y^2 + 7y - 6 \\ 3 \\ \hline \end{array}$ $\quad \begin{array}{r} -4z^2 + 5z + 4 \\ -5 \\ \hline \end{array}$ $\quad \begin{array}{r} 3a^2 - 8a - 5 \\ -6 \\ \hline \end{array}$

5. $\begin{array}{r} 6x^2 - 9x + 1 \\ -4x \\ \hline \end{array}$ $\quad \begin{array}{r} 12y^2 - 4y + 5 \\ 3y \\ \hline \end{array}$ $\quad \begin{array}{r} -2z^2 + 5z - 3 \\ 2z \\ \hline \end{array}$

$-2(3y - 4) = -6y + 8$

$-4(6x^2 - 4x + 8) = -24x^2 + 16x - 32$

6. $5(-6z + 4) =$

$-4(3x^2 - 5) =$

$7(5a^2 + 3a) =$

7. $-3(2z^2 + z - 7) =$

$5(-4n^2 - n + 3) =$

$9(3z^2 + 2z + 4) =$

$3a(5a - 3) = 15a^2 - 9a$

$5y(2y^2 - 4y + 2) = 10y^3 - 20y^2 + 10y$

8. $6n(3n^2 - 8) =$

$9x(3x^2 - 4x) =$

$-2y(4y^2 + 5y) =$

9. $6a(a^2 + 12a - 5) =$

$-2x(3x^2 + 4x - 3) =$

$3s(-2s^2 - 8s + 4) =$

Multiply.

10. $\begin{array}{r} -5a^2 - 4a \\ 3a \\ \hline \end{array}$ $\quad \begin{array}{r} 4x^2 + 6x - 3 \\ -5 \\ \hline \end{array}$ $\quad \begin{array}{r} 9y + 3 \\ 2 \\ \hline \end{array}$ $\quad \begin{array}{r} -4c^2 + 12 \\ 5 \\ \hline \end{array}$

11. $\begin{array}{r} 4x^2 - 5x + 3 \\ 2x \\ \hline \end{array}$ $\quad \begin{array}{r} 14y - 7 \\ -3 \\ \hline \end{array}$ $\quad \begin{array}{r} 7z^2 - 2z \\ z \\ \hline \end{array}$ $\quad \begin{array}{r} -7a^2 - 3a \\ 4a \\ \hline \end{array}$

12. $\begin{array}{r} 7b^2 + 3b \\ 4b \\ \hline \end{array}$ $\quad \begin{array}{r} -5y^2 - 3y + 4 \\ -2y \\ \hline \end{array}$ $\quad \begin{array}{r} 4t^2 - 5t + 3 \\ 6 \\ \hline \end{array}$ $\quad \begin{array}{r} -6s^2 - 8 \\ -7 \\ \hline \end{array}$

MULTIPLYING A BINOMIAL BY A BINOMIAL

To multiply a binomial by a binomial, follow these steps:

Step 1. Multiply each term of the first binomial by each term of the second binomial. Line up like terms.

Step 2. Combine the separate products.

EXAMPLE 1. Multiply $2a - b$ by $3a + 4b$.

Step 1. Multiply $2a - b$ by $4b$.

$4b(2a - b) = 8ab - 4b^2$

Multiply $2a - b$ by $3a$.

$3a(2a - b) = 6a^2 - 3ab$

Step 2. Combine the separate products by adding the columns.

$6a^2 + 8ab - 3ab - 4b^2$

Answer: $= \mathbf{6a^2 + 5ab - 4b^2}$

Multiply:

$$\begin{array}{r} 2a - b \\ \underline{3a + 4b} \\ 8ab - 4b^2 \\ \underline{6a^2 - 3ab } \\ \mathbf{6a^2 + 5ab - 4b^2} \end{array}$$

Multiply.

Examples

$$\begin{array}{r} 5x + 6 \\ \underline{2x + 4} \\ 20x + 24 \\ \underline{10x^2 + 12x } \\ \mathbf{10x^2 + 32x + 24} \end{array}$$

$$\begin{array}{r} 2a + b \\ \underline{a - 2b} \\ -4ab - 2b^2 \\ \underline{2a^2 + ab } \\ \mathbf{2a^2 - 3ab - 2b^2} \end{array}$$

$$\begin{array}{r} 4x - 2y \\ \underline{3x - y} \\ -4xy + 2y^2 \\ \underline{12x^2 - 6xy } \\ \mathbf{12x^2 - 10xy + 2y^2} \end{array}$$

1. $\begin{array}{r} 2y + 3 \\ \underline{y + 5} \end{array}$ $\qquad$ $\begin{array}{r} 3z + 4 \\ \underline{2z + 1} \end{array}$ $\qquad$ $\begin{array}{r} 4a - 6 \\ \underline{3a + 2} \end{array}$

2. $\begin{array}{r} 3x + 4y \\ \underline{2x - y} \end{array}$ $\qquad$ $\begin{array}{r} 7c + 2d \\ \underline{c - 3d} \end{array}$ $\qquad$ $\begin{array}{r} 4r - 3s \\ \underline{2r + 4s} \end{array}$

3. $\begin{array}{r} 2a - 3b \\ \underline{3a - 2b} \end{array}$ $\qquad$ $\begin{array}{r} y - 2z \\ \underline{y - 3z} \end{array}$ $\qquad$ $\begin{array}{r} 5a - 2b \\ \underline{2a - 3b} \end{array}$

Parentheses are often used to indicate multiplication of binomials. Remove parentheses by multiplying each term of the first binomial by each term of the second binomial.

EXAMPLE 2. Multiply: $(4x - 3y)(2x + y)$

$$(4x - 3y)(2x + y) = \underbrace{4x(2x) + 4x(y)}_{\text{Step 1}} + \underbrace{(-3y)(2x) + (-3y)(y)}_{\text{Step 2}}$$

$$= 8x^2 + 4xy - 6xy - 3y^2$$
$$= \mathbf{8x^2 - 2xy - 3y^2}$$

Step 1. Multiply $4x$ by $2x + y$.
$4x(2x + y) = 4x(2x) + 4x(y) = 8x^2 + 4xy$

Step 2. Multiply $-3y$ by $2x + y$.
$-3y(2x + y) = (-3y)(2x) + (-3y)(y) = -6xy - 3y^2$

Step 3. Combine the separate products.
$8x^2 + 4xy - 6xy - 3y^2 = 8x^2 - 2xy - 3y^2$

Answer: $\mathbf{8x^2 - 2xy - 3y^2}$

$(3x + 4)(2x + 1) = 3x(2x) + 3x(1) + 4(2x) + 4(1) = 6x^2 + 3x + 8x + 4$
$= 6x^2 + 11x + 4$

4. $(2y + 3)(4y + 2) =$

$(4a + 2)(2a - 1) =$

$(5z - 3)(3z + 2) =$

$(3m + 2n)(m - n) = 3m(m) + 3m(-n) + 2n(m) + 2n(-n)$
$= 3m^2 - 3mn + 2mn - 2n^2 = \mathbf{3m^2 - mn - 2n^2}$

5. $(4x + 3y)(x - y) =$

$(2a + 3b)(a - 2b) =$

$(3r - 2s)(r + s) =$

6. $(a - b)(a - b) =$

$(3c + 2d)(c + d) =$

$(2y - z)(y - 2z) =$

EXPONENTS IN DIVISION

Terms with like variables can also be divided. As the following examples show, a shortcut to dividing like variables is subtracting exponents.

EXAMPLE 1. Divide y^6 by y^2.

The <u>long way</u> is to write out the numerator and the denominator and then cancel y's.

$$\frac{y^6}{y^2} = \frac{y \cdot y \cdot y \cdot y \cdot y \cdot y}{y \cdot y} = \frac{\cancel{y} \cdot \cancel{y}\ y \cdot y \cdot y \cdot y}{\cancel{y} \cdot \cancel{y}\quad 1} = y^4$$

since there are four y's left in the numerator

Answer: y^4

A <u>*shortcut*</u> is to subtract the exponent of the denominator from the exponent of the numerator as follows:

$$\frac{y^6}{y^2} = y^{6-2} = y^4$$

We can now write the rule for dividing like variables.

<u>RULE:</u> To divide like variables, subtract the exponent of the denominator from the exponent of the numerator.

EXAMPLE 2. Divide z^7 by z^4.

$$\frac{z^7}{z^4} = z^{7-4} = z^3$$

EXAMPLE 3. Divide a^5 by a.

$$\frac{a^5}{a} = a^{5-1} = a^4$$

When the exponent of the denominator is larger than the exponent of the numerator, the answer is shown by putting a 1 over the variable and exponent.

EXAMPLE 4. Divide y^3 by y^5.

$$\frac{y^3}{y^5} = \frac{y \cdot y \cdot y}{y \cdot y \cdot y \cdot y \cdot y} = \frac{\cancel{y} \cdot \cancel{y} \cdot \cancel{y}}{\cancel{y} \cdot \cancel{y} \cdot \cancel{y}} \frac{1}{y \cdot y} = \frac{1}{y^2}$$

since there are 2 y's left in the denominator

We can also solve this problem by subtracting exponents. As you learned in the section on signed numbers, subtracting a larger number from a smaller number gives a negative answer. As Example 4 illustrates, you should write the answer with a positive exponent in the denominator.

$$\frac{y^3}{y^5} = y^{3-5} = y^{-2} = \frac{1}{y^2}$$

Use the method of subtracting exponents to divide the following terms.

Examples

$\frac{x}{x} = x^0 = 1$

$\frac{x^2}{x} = x^1 = x$

$\frac{y^5}{y^3} = y^2$

$\frac{x}{x^3} = x^{-2} = \frac{1}{x^2}$

$\frac{y^3}{y^6} = y^{-3} = \frac{1}{y^3}$

1. $\frac{y}{y} =$ $\frac{z}{z} =$ $\frac{r}{r} =$ $\frac{s}{s} =$ $\frac{t}{t} =$

2. $\frac{z^3}{z} =$ $\frac{y^4}{y} =$ $\frac{a^2}{a} =$ $\frac{b^3}{b} =$ $\frac{r^3}{r} =$

3. $\frac{x^4}{x^2} =$ $\frac{c^6}{c^5} =$ $\frac{b^3}{b} =$ $\frac{c^4}{c^2} =$ $\frac{x^5}{x^3} =$

4. $\frac{a}{a^4} =$ $\frac{y}{y^2} =$ $\frac{x}{x^5} =$ $\frac{z}{z^2} =$ $\frac{b}{b^3} =$

5. $\frac{x^2}{x^4} =$ $\frac{a^4}{a^5} =$ $\frac{b^2}{b^4} =$ $\frac{c^3}{c^6} =$ $\frac{d^5}{d^7} =$

Divide each of the following terms.

6. $\frac{x^3}{x^5} =$ $\frac{a}{a^3} =$ $\frac{t^2}{t} =$ $\frac{c^4}{c^3} =$ $\frac{y}{y} =$ $\frac{z^2}{z^4} =$

7. $\frac{a}{a} =$ $\frac{y^3}{y} =$ $\frac{r^2}{r^2} =$ $\frac{x}{x^3} =$ $\frac{c^4}{c^3} =$ $\frac{s^2}{s^5} =$

8. $\frac{z^3}{z^4} =$ $\frac{x}{x^2} =$ $\frac{y^4}{y^2} =$ $\frac{r^2}{r^3} =$ $\frac{s}{s} =$ $\frac{t^3}{t^6} =$

9. $\frac{a^2}{a^5} =$ $\frac{c^3}{c} =$ $\frac{x^2}{x^3} =$ $\frac{d}{d} =$ $\frac{u^4}{u^5} =$ $\frac{v^3}{v} =$

DIVIDING MONOMIALS

To divide monomials, follow these three steps:

Step 1. Divide the coefficients according to the rules for dividing signed numbers.

Step 2. Divide the variables by subtracting the exponents of each like variable.

Step 3. Write the coefficient and the variables (in alphabetical order).

EXAMPLE 1. Divide $4x^3$ by $-2x^2$.

Step 1. Divide the coefficients: $\frac{4}{-2} = -2$

$$\frac{4x^3}{-2x^2} = \left(\frac{4}{-2}\right) x^{3-2} = \mathbf{-2x}$$

Step 2. Subtract the exponents of the x variable.

$x^{3-2} = x^1 = x$

Step 3. Write the coefficient and variable.

Answer: $\mathbf{-2x}$

EXAMPLE 2. Divide $6a^3b^4c$ by $2a^3b$.

Step 1. Divide the coefficients.

$\frac{6}{2} = 3$

$$\frac{6a^3b^4c}{2a^3b} = \left(\frac{6}{2}\right)a^{3-3}b^{4-1}c = 3a^0b^3c = \mathbf{3b^3c}$$

Step 2. Subtract the exponents of each like variable.

$a^{3-3} = a^0 = 1$ and

$b^{4-1} = b^3$

Step 3. Write the coefficient and variables. $\mathbf{3b^3c}$

Answer: $\mathbf{3b^3c}$

Remember: *Any number to the zero power is 1. Therefore, the letter "a" does not appear in the answer.*

Divide.

Examples

$\frac{9x^4}{3x^2} = 3x^2$

$\frac{14a^4b^2}{2a^2b} = 7a^2b$

1. $\frac{12a^3}{-4a^2} =$ $\qquad$ $\frac{16c^5}{4c^3} =$ $\qquad$ $\frac{-25y^2}{5y} =$

2. $\frac{24u^3v^4}{-6u^2v^3} =$ $\qquad$ $\frac{8d^3e^2}{2d^3e} =$ $\qquad$ $\frac{-35y^5z^3}{5y^2z^2} =$

Examples

$\frac{8a^3b^4}{2a^3b^2} = \mathbf{4b^2}$

$\frac{15x^3y}{6xyz} = \frac{5x^2}{2z}$

$\frac{4x^2}{2x^4} = \frac{2}{x^2}$

$\frac{3a^4b^2}{9a^6b} = \frac{b}{3a^2}$

3. $\frac{-36x^3y^2}{9x^2y^2} =$ $\frac{27c^3d^2}{3cd^2} =$ $\frac{-14xy^3}{-7y^3} =$

4. $\frac{-21ab^3}{14abc} =$ $\frac{21r^3s^3}{7r^2s^2} =$ $\frac{-9xy^4}{6yz} =$

5. $\frac{15a^4}{3a^7} =$ $\frac{-12c^3}{4c^5} =$ $\frac{6d^5}{-2d^6} =$

6. $\frac{-7x^3y}{14x^5y} =$ $\frac{12rs^2}{4r^4s} =$ $\frac{-8uv^3}{-2u^3v^5} =$

Divide the following monomials.

7. $\frac{-6ab^3}{-3ab^2} =$ $\frac{18b^5}{6b^3} =$ $\frac{7u^3v^3}{21uv} =$ $\frac{-9a^3}{-12a} =$

8. $\frac{24x^4y^2}{12xy^2z} =$ $\frac{6a^3b}{-3b^2} =$ $\frac{6y^4}{9y^7} =$ $\frac{13ab^4}{ab} =$

9. $\frac{7xy}{-2xyz} =$ $\frac{8x^2y^4}{12y^5} =$ $\frac{-18c^2d^3}{-21cde} =$ $\frac{3ab^5}{4b^2} =$

10. $\frac{6r^2s^2}{2r^4s^2} =$ $\frac{-36z^5}{9z^2} =$ $\frac{13u^3v^4}{39uv^3} =$ $\frac{7x^3y^4}{21x^3z} =$

11. $\frac{18y^6}{6y^2} =$ $\frac{4xy^3z}{2x^4z} =$ $\frac{9ab^2}{-3b^3} =$ $\frac{12c^3}{-8c^5} =$

12. $\frac{14s^4}{-2s^2} =$ $\frac{15x^2y^3}{5x^2y} =$ $\frac{-9xy^3}{-12xy} =$ $\frac{2abc}{6a^3c^3} =$

DIVIDING A POLYNOMIAL BY A MONOMIAL

To divide a binomial or a trinomial by a monomial, divide each term and combine the separate answers.

EXAMPLE 1. Divide $4y^2 + 8y$ by $2y$.

$$\frac{4y^2 + 8y}{2y} = \frac{4y^2}{2y} + \frac{8y}{2y} = \mathbf{2y + 4}$$

Step 1. Divide $4y^2$ and $8y$ by $2y$.

$$\frac{4y^2}{2y} = 2y \text{ and } \frac{8y}{2y} = 4$$

Step 2. Combine the separate answers.

Answer: $\mathbf{2y + 4}$

EXAMPLE 2. Divide $12a^4b^2 - 6a^2b + 2ab^2$ by $3ab$.

$$\frac{12a^4b^2 - 6a^2b + 2ab^2}{3ab} = \frac{12a^4b^2}{3ab} + \frac{-6a^2b}{3ab} + \frac{2ab^2}{3ab} = \mathbf{4a^3b - 2a + \tfrac{2}{3}b}$$

Step 1. Divide $12a^4b^2$, $-6a^2b$, and $2ab^2$ by $3ab$.

$$\frac{12a^4b^2}{3ab} = 4a^3b \text{ and } \frac{-6a^2b}{3ab} = -2a \text{ and } \frac{2ab^2}{3ab} = \tfrac{2}{3}b$$

Step 2. Combine the separate answers.

Answer: $\mathbf{4a^3b - 2a + \tfrac{2}{3}b}$

Divide.

Examples

$\frac{6x - 4}{2} = 3x - 2$

$\frac{5a^2 + 10a}{5} = a^2 + 2a$

$\frac{2z^2 + 4z}{2z} = z + 2$

$\frac{3z^2 - 6z}{3z^3} = \frac{1}{z} - \frac{2}{z^2}$

1. $\frac{9y + 12}{3} =$ $\qquad$ $\frac{8z - 2}{4} =$

2. $\frac{14x^2 + 7x}{4} =$ $\qquad$ $\frac{6z^2 - 4z}{4} =$

3. $\frac{8a^2 - 6a}{2a} =$ $\qquad$ $\frac{3c^2 + 6c}{3c} =$

4. $\frac{2d^2 + 8d}{2d^4} =$ $\qquad$ $\frac{6u^2 - 3u}{9u^3} =$

Examples

$$\frac{4x^2 + 6x - 8}{2} = 2x^2 + 3x - 4$$

$$\frac{9a^4b^4 - 6a^2b^2 + 3ab}{3ab} = 3a^3b^3 - 2ab + 1$$

5. $\dfrac{6y^2 + 9y - 6}{3} =$

$\dfrac{14z^4 - 7z^2 + 7}{-7} =$

$\dfrac{8a^2 - 6a + 4}{12} =$

6. $\dfrac{12y^2z^2 + 6yz + 2}{2} =$

$\dfrac{-5x^3y^3 + 10x^2y^2 + 5xy}{-5xy} =$

$\dfrac{2u^4v^4 - 4u^2v^2 - 6uv}{2uv} =$

Divide.

7. $\dfrac{10y^2 - 6y}{2y} =$ $\qquad \dfrac{9c - 3}{3} =$ $\qquad \dfrac{4x^3 + 2x}{-2x} =$

8. $\dfrac{4x^4 - x^2}{2x^3} =$ $\qquad \dfrac{5a + 10}{-5} =$ $\qquad \dfrac{14a^2 - 6a}{2} =$

9. $\dfrac{15y^2 - 6y}{3y} =$ $\qquad \dfrac{4z^3 + 3z}{12} =$ $\qquad \dfrac{9u^4 - 8v^3}{8uv} =$

10. $\dfrac{3c^2 + 4c}{4c^3} =$ $\qquad \dfrac{6d^3 + 2d}{4d} =$ $\qquad \dfrac{4x^2 - x}{2x^4} =$

11. $\dfrac{7x^2 - 14x + 21}{7} =$

$\dfrac{3z^3 + 6z + 9}{3} =$

$\dfrac{8u^4 + 4u^2 - 2u}{-2} =$

$\dfrac{7y^2 - 7y + 28}{14} =$

12. $\dfrac{5a^2b^2 - 15ab + 10}{-5} =$

$\dfrac{12c^4d^4 - 6c^2d^2 + 18cd}{6cd} =$

$\dfrac{4x^3y^4 + 6x^2y^2 - 8xy}{4xy} =$

$\dfrac{5xy^4 - 10xy^3 + 15xy^2}{5xy} =$

POLYNOMIALS SKILLS INVENTORY

Add:

1. $\begin{array}{r} +7y \\ \underline{-2y} \end{array}$

2. $\begin{array}{r} -9x^2 \\ \underline{-4x^2} \end{array}$

3. $\begin{array}{r} 6c - 3d \\ \underline{3c - 2d} \end{array}$

4. $\begin{array}{r} 8y^2 - 3y - 6 \\ \underline{3y^2 + 4y + 9} \end{array}$

5. $(-4a + 15b) + (a + 5b) =$

6. $(7y^2 - 6y + 4) + (-3y^2 + 4y - 13) =$

Subtract:

7. $\begin{array}{r} 12z \\ \underline{-9z} \end{array}$

8. $\begin{array}{r} -13a^2 \\ \underline{8a^2} \end{array}$

9. $\begin{array}{r} 4x - 6y \\ \underline{-3x + 5y} \end{array}$

10. $\begin{array}{r} 11c^2 - 9c + 4 \\ \underline{8c^2 - 6c - 5} \end{array}$

11. $(14x - 7y) - (3x + 8y) =$

12. $(8d^2 - 6d + 1) - (3d^2 - 8d - 1) =$

Multiply:

13. $\begin{array}{l} 5y^2 \\ \underline{3\ \ } \end{array}$

14. $\begin{array}{l} z^6 \\ \underline{z^3} \end{array}$

15. $\begin{array}{r} 5ab^2 \\ \underline{-2a^2b} \end{array}$

16. $\begin{array}{r} 5x^2 - 3x \\ \underline{2x} \end{array}$

17. $\begin{array}{r} 6a - 3b \\ \underline{2a - 4b} \end{array}$

18. $-3y(2y^2 - 4y + 11) =$

19. $(-5x - 3)(2x + 6) =$

Divide:

20. $\frac{8z^4}{2} =$ **21.** $\frac{c^4}{c} =$ **22.** $\frac{d^2}{d^5} =$

23. $\frac{-14cd^2}{-2cd} =$ **24.** $\frac{12y^2 - 9y - 15}{-3} =$

25. $\frac{32a^3b^3 - 16a^2b^2 + 24ab}{8ab} =$

POLYNOMIALS SKILLS INVENTORY CHART

Circle the number of any problem that you missed and be sure to review the appropriate practice page. A passing score is 21 correct answers.

Problem Number	Skill Area	Practice Page	Problem Number	Skill Area	Practice Page
1	adding monomials	102	15	multiplying monomials	112
2	adding monomials	102	16	multiplying polynomials and monomials	114
3	adding polynomials	104	17	multiplying binomials	116
4	adding polynomials	104	18	multiplying polynomials and monomials	114
5	adding polynomials	104	19	multiplying binomials	116
6	adding polynomials	104	20	dividing monomials	120
7	subtracting monomials	106	21	exponents in division	118
8	subtracting monomials	106	22	exponents in division	118
9	subtracting polynomials	108	23	dividing monomials	120
10	subtracting polynomials	108	24	dividing polynomials by monomials	122
11	subtracting polynomials	108	25	dividing polynomials by monomials	122
12	subtracting polynomials	108			
13	multiplying monomials and numbers	110			
14	exponents in multiplication	111			

If you missed more than 4 questions, you should review this chapter.

BUILDING NUMBER POWER: ALGEBRA REVIEW TEST

This test will show you how well you have learned all of the skills that you have practiced in this section of the book. Take your time and work each problem carefully.

1. Add: $\begin{array}{r} -7 \\ \underline{+9} \end{array}$

2. $-9 + (-4) =$

3. Subtract: $\begin{array}{r} -14 \\ \underline{+6} \end{array}$

4. $-13 - (-6) =$

5. Multiply: $\begin{array}{r} -13 \\ \underline{7} \end{array}$

6. Divide: $\dfrac{24}{-6}$

7. At 7:00 a.m., the temperature was $-19°$F.
By 11:00 a.m., the temperature was $5°$F.
How many degrees did the temperature rise?

8. Find the value of $(-3)^3$.

9. Simplify $(\frac{2}{3})^3 + (-\frac{1}{3})^2 - (\frac{1}{3})^1$

10. Simplify $\dfrac{5^2\,(-2)^3}{(-4)^3\,5^1}$

11. $\sqrt{49} =$

12. $\sqrt{\dfrac{64}{81}} =$

13. Find an approximate square root of 44.

14. Write an algebraic expression for "Three times the sum of x plus seven."

15. Find the value of $x^3 - y^2 + 5$ for $x = -2$ and $y = 3$.

16. Evaluate the formula $A = \frac{1}{2}bh$ for $b = 6$ feet and $h = 9$ feet. Remember to give your answer in square feet.

17. Solve for x in $\frac{x}{4} - 5 = 3$

18. Solve for y in $2y + 3 = y - 7$

19. Joyce and Vicki run a typing service. As owner of the typewriter, Joyce receives 4 dollars for each 3 dollars received by Vicki. How much would each receive on a job that paid $140?

20. Solve for x in $3x - 2(x - 4) = 7$

21. John, Bill, and Jeff have a house painting business. John earns $40 less a month than Bill. Jeff earns twice as much as John. During a month that the business earned $2,620, how much did each man make?

22. The sum of three consecutive numbers is 87. What is the lowest of the three numbers?

23. Anne earns $900 a month. She pays $225 a month in rent. What is the ratio of her rent to her income?

24. Solve for x in the proportion $\frac{3}{8} = \frac{9}{x}$

25. Marianne earned $225 in 10 days at a temporary job. How much would she earn if she worked for 24 days at the same rate of pay?

26. If 5,280 feet equals 1 mile, what part of a mile is 1,320 feet?

27. Graph the equation $y = -2x + 3$ on the graph at the right for the following values of x:

x	y
-1	
0	
$+1$	

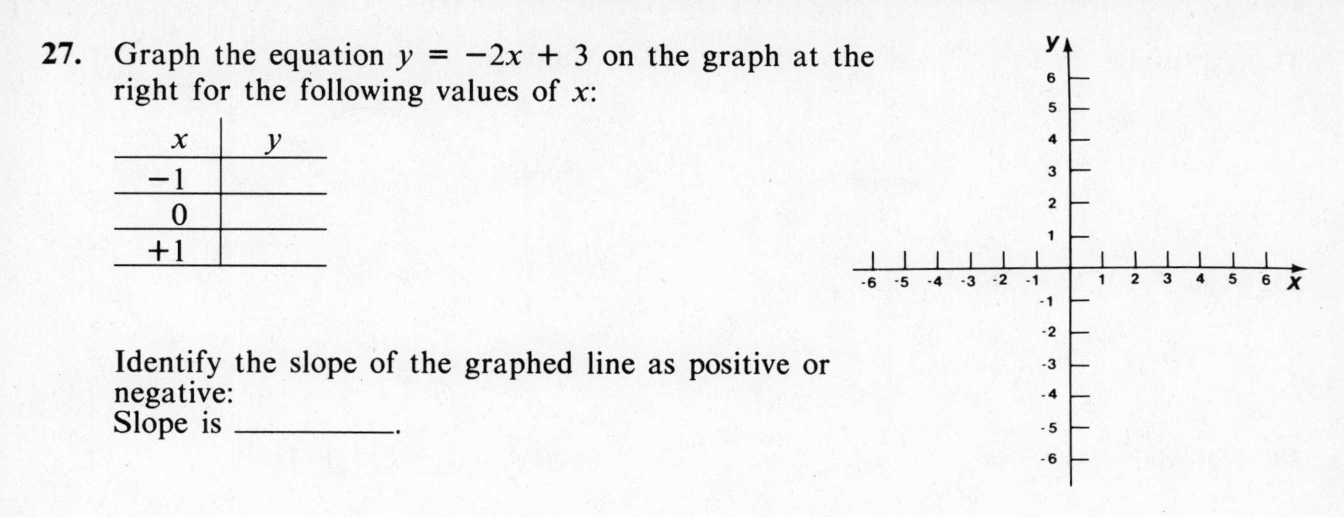

Identify the slope of the graphed line as positive or negative:
Slope is __________.

28. Find the x and y intercepts of a line passing through the following points:

Point A = $(-1,6)$
Point B = $(1,2)$
Point C = $(2,0)$

Do your work on the graph at the right.

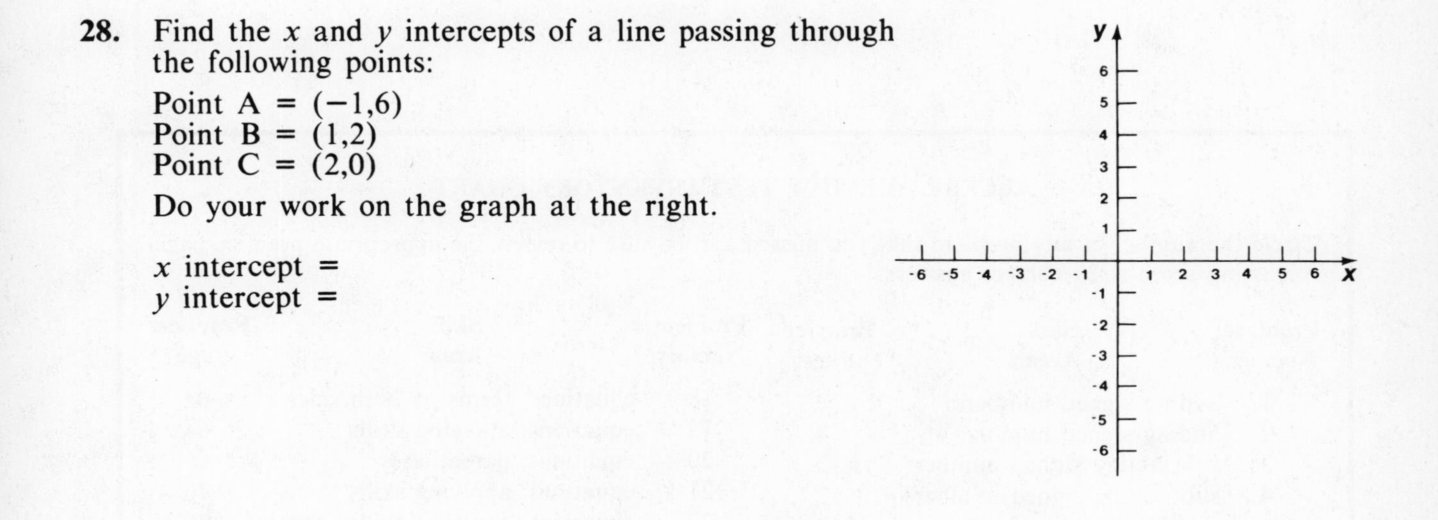

x intercept =
y intercept =

29. $(4x^2 - 6x) + (-3x^2 + 7x) =$

30. $(-4y - 9z) - (6y - 5z) =$

31. Multiply: $\begin{array}{r} 2x^3y \\ \underline{-xy^2} \end{array}$

32. Multiply: $\begin{array}{r} 3x^2 + 2x \\ \underline{\qquad 6} \end{array}$

33. $(3a + b)(2a - 2b) =$

34. Divide: $\dfrac{-9a^3b^2}{a^2b} =$

35. Divide: $\dfrac{6x^3 - 8x^2 + 4x}{-2x}$

ALGEBRA REVIEW TEST INVENTORY CHART

Circle the number of any problem that you missed and be sure to review the appropriate practice page. A passing score is 29 correct answers.

Problem Number	Skill Area	Practice Page	Problem Number	Skill Area	Practice Page
1	adding signed numbers	8	18	equations: terms on both sides	62
2	adding signed numbers	8	19	equations: applying skills	64
3	subtracting signed numbers	12	20	equations: parentheses	66
4	subtracting signed numbers	12	21	equations: applying skills	70
5	multiplying signed numbers	14	22	equations: applying skills	70
6	dividing signed numbers	16	23	ratio: applying skills	73
7	signed numbers: applying skills	20	24	proportion	74
			25	proportion: applying skills	76
8	powers	24	26	proportion: applying skills	76
9	powers	26	27	slope and graphing	92 and 94
10	powers: multiplication and division	27	28	finding intercepts	90
			29	adding polynomials	104
11	square root	28	30	subtracting polynomials	108
12	square root	29	31	multiplying monomials	112
13	square roots: applying skills	30	32	multiplying monomials and numbers	110
14	writing algebraic expressions	34	33	multiplying binomials	116
15	evaluating algebraic expressions	37	34	dividing monomials	120
16	evaluating formulas	39	35	dividing polynomials by monomials	122
17	equations: several operations	56			

Review any remaining problem areas. If you passed the test, go on to Using Number Power. If you did not pass the test, take the time for a more thorough review of the book.

USING NUMBER POWER

WIND CHILL

The temperature that is shown on a thermometer is the temperature that you feel when there is no wind. The term *wind chill* means that the temperature together with the wind causes you to feel colder than the thermometer reads. For example, a wind speed of 25 miles per hour at a temperature of 10°F feels as cold as −29°F with no wind blowing.

The table below shows thermometer temperatures with the equivalent temperatures when the wind blows at different speeds. The example given above is indicated on the table by arrows.

Wind speed (mph)	Thermometer temperature (°F)										
	−25	−20	−15	−10	−5	0	5	10	15	20	25
	Equivalent temperature with the wind blowing (°F)										
5	−31	−26	−21	−15	−10	− 5	0	6	11	16	22
10	−52	−46	−40	−34	−27	−22	−15	− 9	− 3	3	10
15	−65	−58	−51	−45	−38	−31	−25	−18	−11	− 5	2
20	−74	−67	−60	−53	−46	−39	−31	−24	−17	−10	− 3
25	−81	−74	−66	−59	−51	−44	−36	−29	−22	−15	− 7
30	−86	−79	−71	−64	−56	−49	−41	−33	−25	−18	−10
35	−89	−82	−74	−67	−58	−52	−43	−35	−27	−20	−12
40	−92	−84	−76	−69	−60	−53	−45	−37	−29	−21	−13

EXAMPLE: When the wind speed is 20 mph, the wind chill temperature is how many degrees below the thermometer temperature of 10°F?

Step 1. On the table above, find the equivalent temperature of 10°F when the wind blows 20 mph: −24°F

Step 2. Find the difference between the temperatures by subtracting. Use the rules for subtracting signed numbers.
thermometer temperature − wind chill temperature = difference
10 − (−24) = 10 + 24 = **34**

Answer: The wind chill temperature is **34 degrees** below the thermometer temperature.

Using the chart above, find the difference between the thermometer temperature and the wind chill temperature.

1. Thermometer temperature = 15°F, wind speed = 10 mph

2. Thermometer temperature = −10°F, wind speed = 15 mph

3. Thermometer temperature = 20°F, wind speed = 35 mph

4. Thermometer temperature = 0°F, wind speed = 30 mph

5. Thermometer temperature = −15°F, wind speed = 5 mph

6. Thermometer temperature = 10°F, wind speed = 40 mph

7. Which has the greater difference between thermometer and wind chill temperatures: wind speed 15 mph at temperature −10°F or wind speed 20 mph at a temperature of 25°F?

8. Which combination of temperature and wind would make you feel colder: wind speed of 15 mph at −15°F or wind speed of 25 mph at a temperature of 0°F?

THE DISTANCE FORMULA

The distance formula $D = rt$ tells how to find the distance traveled (D) when you know the rate (r)—or speed—and the time of travel (t).

EXAMPLE 1. How far can Joan drive in 7 hours if she averages 50 miles per hour?

Step 1. Identify r and t:
$r = 50$ miles per hour and $t = 7$ hours

Step 2. Substitute the r and t values into the distance formula.
$D = rt = (50)(7) =$ **350 miles**

Answer: **350 miles**

The distance formula can be rewritten in two other ways: as a rate formula and as a time formula.

(1) As a rate formula $r = \frac{D}{t}$

The rate formula is used to find the rate—or speed—when the distance traveled and time of travel are known.

EXAMPLE 2. On part of her trip, Joan drove 405 miles in 9 hours. What was her average speed?

Step 1. Identify D and t:
$D = 405$ miles and $t = 9$ hours

Step 2. Substitute the D and t values into the rate formula and solve.
$r = \frac{D}{t} = \frac{405}{9} =$ **45 miles per hour**

Answer: **45 miles per hour**

(2) As a time formula $t = \frac{D}{r}$

The time formula is used to find the time of travel when the distance traveled and rate—or speed—are known.

EXAMPLE 3. On the last day of her trip, Joan had 280 miles left to go. If she averaged 40 miles per hour, how long would it take her to drive this distance?

Step 1. Identify D and r:
$D = 280$ miles and $r = 40$ miles per hour

Step 2. Substitute the D and r values into the time formula.
$t = \frac{D}{r} = \frac{280}{40} =$ **7 hours**

Answer: **7 hours**

In the following problems, decide whether you are looking for the distance (*D*), rate (*r*)—or speed, or time of travel (*t*). Then use the appropriate formula to solve.

1. What distance can Bert drive in 12 hours if he averages 50 miles per hour?

2. On the second day of his trip, Bert drove 440 miles in 8 hours. Find his average speed during the second day.

3. If Bert could average 50 miles per hour on the third day, how long would it take him to drive 350 miles?

4. Jennifer drove from Chicago to New York. On the first day of her trip, she averaged 48 miles per hour for 6 hours. How far did she drive the first day?

5. On the second day of her trip, Jennifer drove 364 miles in 7 hours. Find her average speed during the second day.

6. On the third day, Jennifer averaged 54 miles per hour as she drove 324 miles. How many hours did Jennifer drive on the third day?

SIMPLE INTEREST FORMULA

Interest is money that is earned (or paid) for the use of money. If you deposit money in a savings account, interest is the money that the bank pays you for using your money. If you borrow money, interest is the money that you pay for using the lender's money.

Simple interest is interest on an unchanging principal only (the amount that is deposited or borrowed).

The *simple interest formula* is $I = PRT$.

I = interest, written in dollars
P = principal, money deposited or borrowed, written in dollars
R = percentage rate, written as a fraction or decimal
T = time, written in years

Read $I = PRT$ as "Interest equals principal times rate times time."

EXAMPLE 1. What is the interest earned on $300 deposited for 2 years in a savings account which pays 5% simple interest?

Step 1. Identify P, R, and T.

$P = 300 \qquad R = 5\% = \frac{5}{100} \qquad T = 2$

The interest rate, 5%, is expressed as the fraction $\frac{5}{100}$ to simplify multiplication in Step 2.

Step 2. Substitute and multiply. Cancel if you can.

$I = PRT = {}^{3}\cancel{300}\left(\frac{5}{\cancel{100}_1}\right) 2 = \mathbf{\$30}$

Answer: **$30**

EXAMPLE 2. At the end of 3 years, what is the total amount owed on a $429 loan borrowed at 16½% simple interest per year?

Step 1. Identify P, R, and T.

$P = 429 \qquad R = 16\frac{1}{2}\% = .165 \qquad T = 3$

It is sometimes easier to multiply with decimals than with fractions, especially when there would be little or no cancellation with fractions.

Step 2. Substitute and multiply.

```
   429     P             70.785   PR
 ×.165    ×R                 ×3   ×T
  2145                  212.355   PRT
 2574                = $212.36   The interest (PRT) is rounded to
 429                             the nearest cent.
70.785    PR
```

Step 3. Principal + Interest = Total amount owed

$429 + 212.36 = **$641.36**

Answer: **$641.36**

In the following problems, decide whether you are looking for the interest only or for the new principal (the original principal plus the interest) and solve.

1. What is the interest earned on $200 deposited for 3 years in a savings account that pays 5% simple interest?

2. Jane deposited $650 at a 5½% interest rate. What would be the total amount in her account after 2 years?

3. At a 10% interest rate, how much interest would Bill have to pay on $475 borrowed for 2 years?

4. What is the total amount owed on a loan if the principal is $1,000, the interest rate is 14%, and the time is 3 years?

5. How much interest would John pay for a loan of $575 at 12½% if he repaid the bank at the end of 1 year?

6. What is the total amount now in a savings account if the amount deposited was $375, the interest rate was 5¼%, and the time was 4 years?

SIMPLE INTEREST FORMULA: PARTS OF A YEAR

Interest is paid on the basis of a yearly percentage rate. However, not all deposits or loans are made for whole years. To use the simple interest formula for a time period that is not a whole year, write the time as a fraction of a year.

EXAMPLE 1. How much interest is earned on a $600 deposit at an interest rate of 7% for 8 months?

Step 1. Identify P, R, and T.

$P = 600 \quad R = \frac{7}{100} \quad T = \frac{8}{12} = \frac{2}{3}$

To write 8 months as a fraction of a year, write 8 over 12—the number of months in a year. The fraction $\frac{8}{12}$ can then be reduced to $\frac{2}{3}$.

Step 2. Substitute and multiply. Cancel if possible.

$\text{I} = PRT = {}^{6}\cancel{600}\left(\frac{7}{\cancel{100}_1}\right)\frac{2}{3} = {}^{2}\cancel{6}(7)\frac{2}{\cancel{3}_1} = \$28$

Answer: **$28**

EXAMPLE 2. What is the interest paid on a loan of $800 borrowed for 2 years and 3 months at 9% simple interest?

Step 1. Identify P, R, and T.

$P = 800 \quad R = \frac{9}{100} \quad T = 2\frac{3}{12} = 2\frac{1}{4} = \frac{9}{4}$

Notice that in the mixed number, you write 2 for 2 whole years and $\frac{3}{12}$ or $\frac{1}{4}$ for the months. Write the time as an improper fraction to simplify multiplication in Step 2.

Step 2. Substitute and multiply. Cancel if possible.

$\text{I} = PRT = {}^{8}\cancel{800}\left(\frac{9}{\cancel{100}_1}\right)\frac{9}{4} = {}^{2}\cancel{8}(9)\frac{9}{\cancel{4}_1} = \$162$

Answer: **$162**

EXAMPLE 3. What is the interest earned on $200 deposited for 16 months in a savings account paying 6% interest?

Step 1. Identify P, R, and T.

$P = 200 \quad R = \frac{6}{100} \quad T = \frac{16}{12} = \frac{4}{3}$

Step 2. Substitute and multiply. Cancel if possible.

$\text{I} = PRT = {}^{2}\cancel{200}\left(\frac{{}^{2}\cancel{6}}{\cancel{100}_1}\right)\frac{4}{\cancel{3}_1} = \$16$

Answer: **$16**

In the following problems, decide whether you are looking for the interest or the new principal (original principal plus the interest) and solve.

1. What is the interest earned on $900 deposited for 1 year 6 months in a savings account that pays 5% interest?

2. How much money will Tracy repay to the bank for a loan of $1,000 that she borrowed for 2 years 8 months at a 12% interest rate?

3. Eva deposited $850 in a savings account. How much interest will her money earn after 20 months if the interest rate is 6%?

4. How much money will George have to repay to the bank if he borrowed $1,000 for 9 months at an interest rate of 15%?

5. New Frontier Furniture Company charged Anne 18% interest on her $750 purchase. How much interest must Anne pay if she pays the entire bill after 10 months?

6. At a 5% interest rate, how much interest would $1,200 earn if deposited for 14 months?

COMPOUND INTEREST FORMULA

Did you ever wonder how banks calculate interest? They actually pay you more than simple interest. They pay *compound interest.*

Compound interest works this way: a bank pays simple interest at the end of a short time period, and then that interest is added to the principal. During the next time period, the added interest also earns interest. The interest earned is called *compound interest*, and the process is called *compounding interest*.

To find the value of money earning compound interest, we use the *compound interest formula:*

$$A = P(1 + \frac{R}{m})^n$$

A = Accumulated amount: original principal plus all interest earned
P = Original principal
R = Yearly percentage rate
m = Number of time periods per year that interest is paid
n = Total number of time periods that interest is paid (m times the number of years)

EXAMPLE: If $5,000 is placed in a savings account earning 6% interest compounded every 4 months, what will be the total amount in savings at the end of 1 year?

Step 1. Identify P, R, m, and n.
In compound interest problems, R is always written as a decimal.
To find m, divide 1 year (12 months) by the pay period—4 months.
To find n, multiply m (3) by the number of years (1): 3(1) = 3
$P = 5{,}000$ $R = 6\% = .06$ $m = \frac{12}{4} = 3$ $n = 3(1) = 3$

Step 2. The total amount in savings is represented as A in the formula. Find A as follows:

a) Substitute values in the formula.
$A = P(1 + \frac{R}{m})^n$ $= 5{,}000(1 + \frac{.06}{3})^3$

b) Evaluate the parentheses.
$\frac{.06}{3} = .02$ $= 5{,}000(1 + .02)^3$
$1 + .02 = 1.02$ $= 5{,}000(1.02)^3$

c) Evaluate the power.
$(1.02)^3 = 1.061208$ $= 5{,}000(1.061208)$

d) Multiply.
$5000(1.061208) = 5306.040$ $= 5{,}306.040$
Round off the answer to the nearest cent (hundredth). = **$5,306.04**

Note: If you worked the problem as a simple interest problem for 1 year, the answer would be $5,300.

Answer: **$5,306.04**

Solve the problems below.

1. If $2,000 is placed in a savings account earning 6% interest that is compounded every 4 months, what will be the total amount in savings at the end of 1 year?

2. If $1,000 is placed in a savings account earning 6% interest that is compounded every 6 months, what will be the total amount in savings at the end of 2 years?

3. If $500 is invested and earns 10%, compounded every 3 months, what will the investment be worth at the end of 1 year?

WORK PROBLEMS

A common problem—often called a *work problem*—involves finding the rate at which 2 or more people working together can do a job. If you know the rate at which each person works alone, you can find the rate at which they work together.

EXAMPLE: John can do a yard job alone in 2 hours. Bill can do the same job alone in 3 hours. How long will it take the two of them working together to finish the job?

Step 1. Express the fraction of the job each can do in 1 hour.

$\frac{1}{2}$ = fraction of the job John can do in 1 hour.

(If he can do the whole job in 2 hours, he can do $\frac{1}{2}$ the job in 1 hour.)

$\frac{1}{3}$ = fraction of the job Bill can do in 1 hour.

Step 2. Express the fraction of the job each can do in x hours, when x = the time it takes to complete the job together.

$\frac{1}{2}x$ = fraction of the job John can do in x hours.

$\frac{1}{3}x$ = fraction of the job Bill can do in x hours.

Step 3. Set the sum of the two fractions equal to 1 and solve the resulting equation for x.

Note: Step 3 says that the sum of the parts of the job (the fractions) is equal to the whole job, represented by the number 1.

$\frac{1}{2}x + \frac{1}{3}x = 1$ Add the x's.

$\frac{5}{6}x = 1$ since $\frac{1}{2} + \frac{1}{3} = \frac{5}{6}$

$\left(\frac{\not{6}^1}{\not{5}_1}\right)\frac{\not{5}^1}{\not{6}_1}x = 1\left(\frac{6}{5}\right)$ Multiply by the reciprocal of $\frac{5}{6}$ so that the x stands alone.

$x = \frac{6}{5} = \mathbf{1\frac{1}{5}}$ **hours**

Answer: $\mathbf{1\frac{1}{5}}$ **hours**

The steps used to solve the example can be summarized as follows:

Step 1. Find the fractional part of the job each can do in one hour.

Step 2. Multiply each fraction from Step 1 by x.

Step 3. Set the sum of the fractions from Step 2 equal to the number 1 and solve for x.

Solve the problems below.

1. Working alone, Carrie can paint a room in 4 hours. Amy, working alone, can paint the same room in 3 hours. How long will it take the two of them, working together, to paint the room?

2. George can wax a large truck in 5 hours when he is working alone. Greg can do the same job alone in 4 hours. How long will it take George and Greg to wax the truck if they work together?

3. By herself, Mary can trim the fruit trees in 2 hours. Bill can do the same job by himself in 4 hours. Working together, how long will it take Bill and Mary to trim the trees?

4. Using the larger tractor, Don can plow a field in 3 hours. Using the smaller tractor, he could plow the same field in 5 hours. If he has a friend drive the smaller tractor, how long will it take them to plow the field working together?

5. Anne can wash the windows in 40 minutes if she works alone. Jeff, working alone, takes an hour to do the same windows. If they work together, how long will it take them to do the windows? (*Hint:* 40 minutes is what part of an hour?)

MIXTURE PROBLEMS: PART I

Did you ever wonder how the prices are determined for different mixtures of nuts or candies? For example, you know that a mixture of cashews and peanuts costs more than peanuts alone but costs less than cashews alone. How is the exact price determined? With algebra, you can answer this question yourself.

EXAMPLE: How many pounds of a \$5 per pound mixture of nuts must be mixed with 3 pounds of a \$2 per pound mixture of nuts to obtain a new mixture with a value of \$4 per pound?

Step 1. Let x = the amount of \$5 per pound mixture of nuts needed.
Now, put the information you know in a table as follows:

a) Row I is the Cost per Pound of each mixture.
b) Row II is the Number of Pounds of each mixture.
c) Row III is the dollar value of each mixture: Value (\$).
Value (\$) = Cost per Pound times Number of Pounds

Row I	Cost per Pound:	\$5	\$2	\$4
Row II	Number of Pounds:	x	3	$x + 3$
Row III	Value (\$):	$5x +$	$6 =$	$4(x + 3)$

Row III gives the equation you are to solve. In words, the value of the \$5 per pound mixture ($5x$) plus the value of the \$2 per pound mixture ($2 \cdot 3 = 6$) is equal to the value of the \$4 per pound mixture $4(x + 3)$:
In words: \$5 mixture + \$2 mixture = \$4 mixture
In symbols: $5x + 6 = 4(x + 3)$

Step 2. Solve the equation for x.

$5x + 6 = 4(x + 3)$

$5x + 6 = 4x + 12$ — Remove the parentheses

$5x = 4x + 12 - 6$ — Subtract the 6 from the 12

$5x = 4x + 6$

$5x - 4x = 6$ — Subtract the $4x$ from the $5x$

$\mathbf{x = 6}$

Answer: **x = 6 pounds:** 6 pounds of the \$5 per pound mixture must be used.

Solve the problems below.

1. How many pounds of a $6 per pound mixture of nuts must be mixed with 5 pounds of a $3 per pound mixture of nuts to obtain a new mixture worth $5 per pound?

2. How many pounds of a dried fruit mix costing $3.00 per pound must be mixed with 8 pounds of a nut mixture costing $4.50 per pound to give a fruit-nut mixture costing $4.00 per pound?

3. How many pounds of chocolates costing $2.00 per pound should be mixed with 3 pounds of chocolates costing $1.60 per pound to obtain a mixture of chocolates costing $1.70 per pound?

MIXTURE PROBLEMS: PART II

Some mixture problems involve mixing liquids of different strengths. The new mixture has a strength that depends on the amount and strength of each liquid used in the mixture.

Study the following example before beginning the problems.

EXAMPLE: How many gallons of a liquid that is 70% alcohol must be mixed with 5 gallons of one that is 40% alcohol to obtain a mixture that is 50% alcohol?

Step 1. Let x = the amount of liquid needed that is 70% alcohol.
Look at the following table of information.

		First Liquid		Second Liquid		Mixture
Row I	Percent	70% = .70		40% = .40		50% = .50
Row II	Number of Gallons	x		5		$x + 5$
Row III	Number of Gallons of Alcohol	$.70x$	+	2	=	$.50(x + 5)$

Since the number of gallons of alcohol in the mixture is equal to the sum of the number of gallons in the two liquids, you can write an equation expressing these equal total values.

$.70x + 2 = .50(x + 5)$, which can be written as:

$.7x + 2 = .5(x + 5)$

Step 2. Solve the equation for x.

$.7x + 2 = .5(5 + x)$

$.7x + 2 = 2.5 + .5x$

$.7x = 2.5 - 2 + .5x$ Subtract 2 from 2.5.

$.7x = .5 + .5x$

$.7x - .5x = .5$ Subtract .5x from .7x.

$.2x = .5$

$x = \frac{.5}{.2}$ Divide .5 by .2.

$x = 2.5$

Answer: x **= 2.5 gallons.**

Solve the problems below.

1. How many gallons of a liquid that is 40% alcohol must be mixed with 8 gallons of one that is 30% alcohol to obtain a mixture that is 35% alcohol?

2. How many pounds of concrete that is 9% cement must be mixed with 200 pounds of concrete that is 12% cement to obtain a concrete mixture that is 10% cement?

3. How many quarts of whipping cream that is 36% butterfat must be mixed with 4 quarts of half-and-half that is 12% butterfat to make light cream that is 18% butterfat?

TEMPERATURE: FINDING CELSIUS FROM FAHRENHEIT

Temperature is measured in units called "degrees." The United States currently uses the *Fahrenheit* scale thermometer, while most other countries in the world use the *Celsius* scale thermometer. The Celsius scale is often called the *Centigrade* scale.

A line graph is a good way to see the relationship between the Fahrenheit and Celsius scales.

On the line graph at the right, equivalent temperatures are plotted and the points are connected by a straight line.

"Degrees Fahrenheit"—read on the horizontal axis—is abbreviated "°F."

"Degrees Celsius"—read on the vertical axis—is abbreviated "°C."

The coordinates of a point on the line give equivalent Fahrenheit and Celsius temperatures. A few points have been identified as examples:

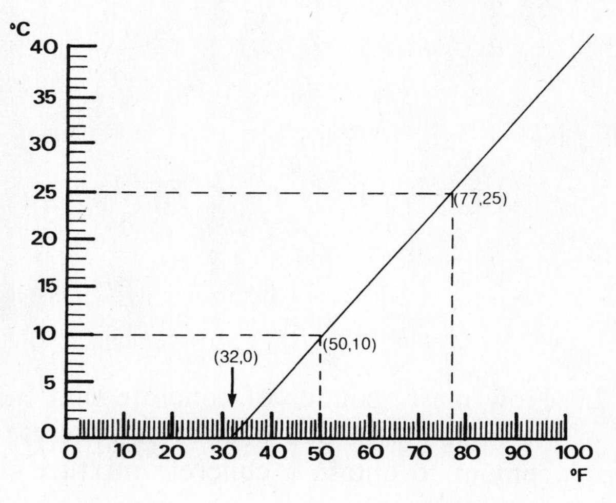

EXAMPLES: 32°F = 0°C
50°F = 10°C
77°F = 25°C

Refer to the graph above and answer each question with either "Fahrenheit" or "Celsius."

1. Which temperature scale is shown on the horizontal coordinate axis? ____________
2. Which temperature scale is shown on the vertical coordinate axis? ____________
3. Which temperatures are represented along the "x number line"? ____________
4. Which temperatures are represented along the "y number line"? ____________

From the graphed line, find the approximate equivalent Celsius temperature for each Fahrenheit temperature listed below:

5. 32°F = 6. 59°F = 7. 41°F =

8. 68°F = 9. 95°F = 10. 86°F =

Answer each question about the graphed temperature line.

11. What temperature value is at the "x intercept"? ____________
12. Is the slope of the temperature line positive or negative? ____________

The graph on the previous page is useful for quickly finding approximate equivalent temperatures. To accurately find the Celsius temperature when you know the Fahrenheit temperature, use the equation that relates the two temperature scales:

$$°C = \frac{5}{9}(°F - 32°)$$

EXAMPLE: Find the Celsius temperature when the Fahrenheit temperature is 98.6°. $°C = \frac{5}{9}(°F - 32°)$

Step 1. Substitute 98.6 for $°F$ $°C = \frac{5}{9}(98.6 - 32)$

Step 2. Evaluate the parentheses.
$(98.6 - 32) = (66.6)$ $°C = \frac{5}{9}(66.6)$

Step 3. Multiply by $\frac{5}{9}$:
Multiply by 5: $5(66.6) = 333$ $°C = \frac{333}{9}$
Divide by 9: $333 \div 9 =$ **37°** $°C =$ **37°**

Answer: 37°C

Use the temperature equation above to answer the following questions.

13. Jim set the room thermostat to 68°F to help save energy. What was the equivalent Celsius temperature?

14. What is the Celsius temperature of an oven set at 410°F to bake a cake?

15. The normal human body temperature is 98.6°F, which is 37°C. When Sally was ill, her temperature was 104°F. What was her temperature in °C?

16. The boiling point of water is 212°F. What is the equivalent Celsius temperature for the boiling point of water?

TEMPERATURE: FINDING FAHRENHEIT FROM CELSIUS

The temperature equation introduced on page 148 can be rewritten without parentheses so that °F appears on the left side of the equal sign:

$$°F = \frac{9}{5}°C + 32°$$

In this form, the equation can be used to find an equivalent Fahrenheit temperature for a given Celsius temperature.

EXAMPLE:	Find the Fahrenheit temperature when the Celsius temperature is 45°.	$°F = \frac{9}{5}°C + 32°$
Step 1.	Substitute 45° for *°C*.	$°F = \frac{9}{5}(45) + 32$
Step 2.	Multiply 45 by $\frac{9}{5}$.	
	Multiply 45 by 9: $45 \times 9 = 405$	$°F = \frac{405}{5} + 32$
	Divide 405 by 5: $405 \div 5 = 81$	
Step 3.	Add.	$°F = 81 + 32$
Answer:	**113°F**	$\mathbf{°F = 113°}$

The "Tables of Values," the "Points to Graph," and the blank graph at the lower right will be used as you work on the following problems.

Table of Values

°C	°F
15	
25	
35	

Points to Graph

(,)
(,)
(,)

1. Label the x coordinate axis (horizontal axis) as "Celsius Temperatures" or "°C."
2. Label the y coordinate axis (vertical axis) as "Fahrenheit Temperatures" or "°F."
3. Use the temperature equation to find Fahrenheit temperatures equivalent to the Celsius temperatures given in the Tables of Values. Write your answers in the table, and fill in the Points to Graph column with the coordinates of the points from the table.
4. Plot these points (from 3.) on the graph.
5. Connect the points with a straight line, and extend the line to the edges of the graph.

Use the temperature equation on the previous page to answer the following questions.

6. Water freezes at 0° Celsius. What is the Fahrenheit temperature at which water freezes?
7. Mary likes to be outdoors when the temperature reaches 25°C. What would be the equivalent Fahrenheit temperature?
8. What is the oven temperature in °F when meat is broiling at 290°C?
9. When Raul had the flu, his temperature was 38°C. What was his temperature in °F?
10. 20°C is a comfortable room temperature. What is the equivalent temperature in °F?

USING ALGEBRA IN GEOMETRY

Algebra is often used to solve problems in geometry. In the two examples below, unknown angles are represented by variables (letters), and equations are written and solved to find the size of each angle.

As each example is discussed, some ideas from geometry will be introduced.

In geometry, an angle (∠) is measured in units called "degrees" (°). A right angle (∟), formed by perpendicular lines contains 90°. An angle can be labeled with three letters, the second letter being the point where the two sides of the angle meet.

EXAMPLE 1. ∠ ABC is a right angle (90°) that is divided into three smaller angles as shown in the drawing:
∠ DBE = 2 times ∠ EBC *or* 2 ∠ EBC and
∠ ABD = 3 times ∠ EBC *or* 3 ∠ EBC.
How large is each of the smaller angles? (in degrees)

Step 1. Let ∠ EBC = x, ∠ DBE = $2x$, and ∠ ABD = $3x$

Step 2. Set the sum of the angles equal to 90° (the total number of degrees) and solve the equation.

$$x + 2x + 3x = 90$$
$$6x = 90$$
$$\mathbf{x} = \frac{90}{6} = \mathbf{15}$$
$$\mathbf{2x} = 2(15) = \mathbf{30}$$
$$\mathbf{3x} = 3(15) = \mathbf{45}$$

Answer: **∠ EBC = 15°, ∠ DBE = 30°,** and **∠ ABD = 45°**

For Example 2, you need to know that the sum of the three angles in a triangle always equals 180°. Also, an angle in a triangle is often represented by one letter only.

EXAMPLE 2. In the triangle at the right, ∠ B = 3 ∠ A, and ∠ C = 5 ∠ A. How many degrees is each angle?

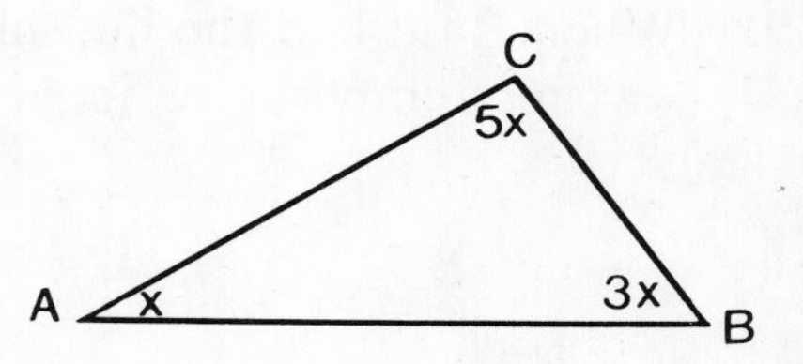

Step 1. Let ∠ A = x, ∠ B = $3x$, and ∠ C = $5x$.

Step 2. Set the sum of the angles equal to 180°, and solve the equation.

$$x + 3x + 5x = 180$$
$$9x = 180$$
$$\mathbf{x} = \frac{180}{9} = \mathbf{20}$$
$$\mathbf{3x} = 3(20) = \mathbf{60}$$
$$\mathbf{5x} = 5(20) = \mathbf{100}$$

Answer: ∠ **A = 20°,** ∠ **B = 60°,** and ∠ **C = 100°**

In each problem below, solve for the angles as indicated.

1. $\angle ABC = 90°$
$\angle DBE = 3\angle EBC$
$\angle ABD = 5\angle EBC$

How many degrees are in each of these angles?
$\angle EBC =$

$\angle DBE =$

$\angle ABD =$

2. $\angle A + \angle B + \angle C = 180°$
$\angle B = 2\angle A$
$\angle C = 3\angle A$

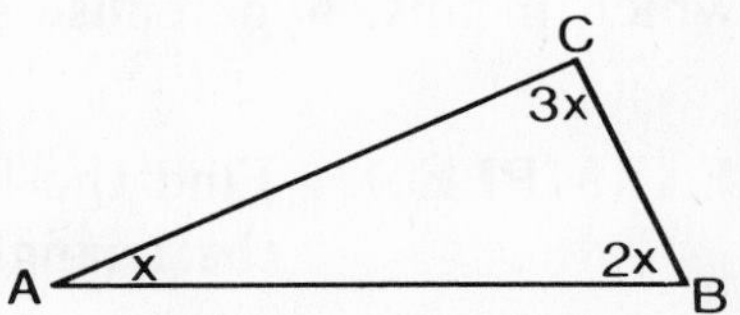

How many degrees are in each of these angles?
$\angle A =$

$\angle B =$

$\angle C =$

3. $\angle ABC = 90°$
$\angle DBE = 5\angle EBC$
$\angle ABD = 9\angle EBC$

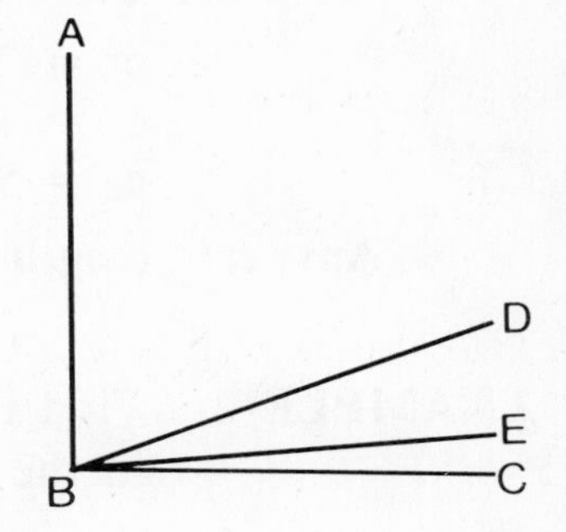

How many degrees are in each of these angles?
$\angle EBC =$

$\angle DBE =$

$\angle ABD =$

4. $\angle A + \angle B + \angle C = 180°$
$\angle B = 5\angle A$
$\angle C = 4\angle A$

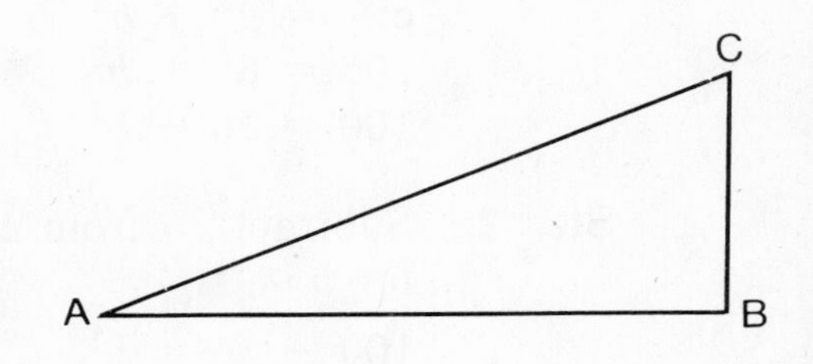

How many degrees are in each of these angles?
$\angle A =$

$\angle B =$

$\angle C =$

RIGHT TRIANGLES AND THE PYTHAGOREAN THEOREM

A *right triangle* is a triangle in which two sides meet at a right angle — perpendicular to one another. The side opposite the right angle is called the *hypotenuse*.

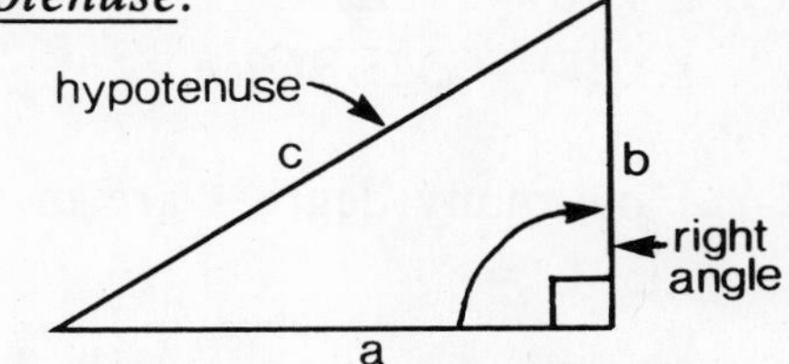

The Greek mathematician Pythagoras discovered that the square of the length of the hypotenuse of a right triangle is equal to the sum of the squares of the lengths of the other two sides. We call this relation the *Pythagorean Theorem*.

In symbols, using the labels on the triangle above, we can write the Pythagorean Theorem as:

$$\text{Pythagorean Theorem:} \quad c^2 = a^2 + b^2$$

which means: hypotenuse squared = side squared + side squared.

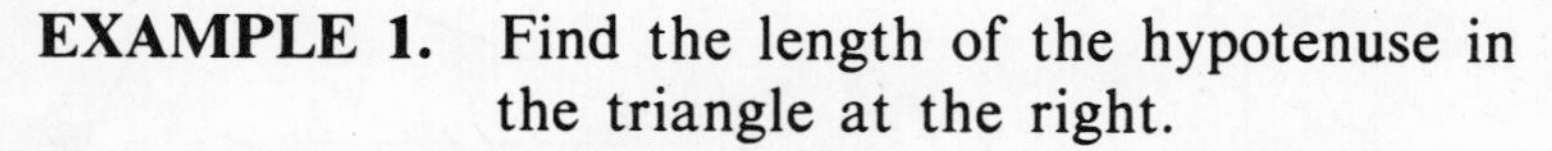

EXAMPLE 1. Find the length of the hypotenuse in the triangle at the right.

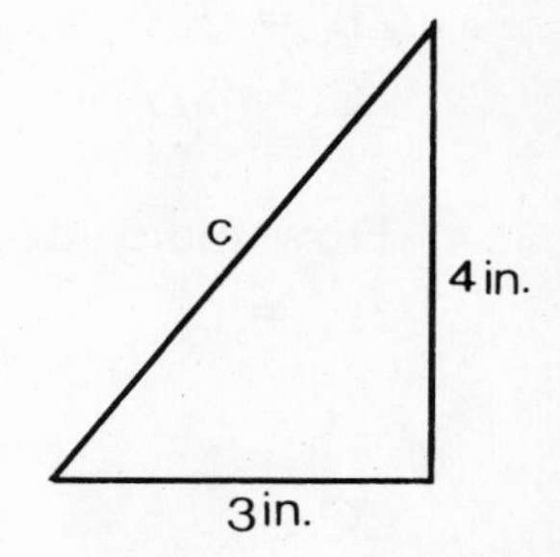

Step 1. Substitute 3 for a and 4 for b in the Pythagorean Theorem.

$c^2 = a^2 + b^2$
$c^2 = 3^2 + 4^2$
$c^2 = 9 + 16$
$c^2 = 25$

Step 2. Solve for c.

$c^2 = 25$
$c = \sqrt{25}$
$c = 5$

Answer: **length of hypotenuse = 5 inches**

EXAMPLE 2. Find the length of the unlabeled side in the triangle to the right.

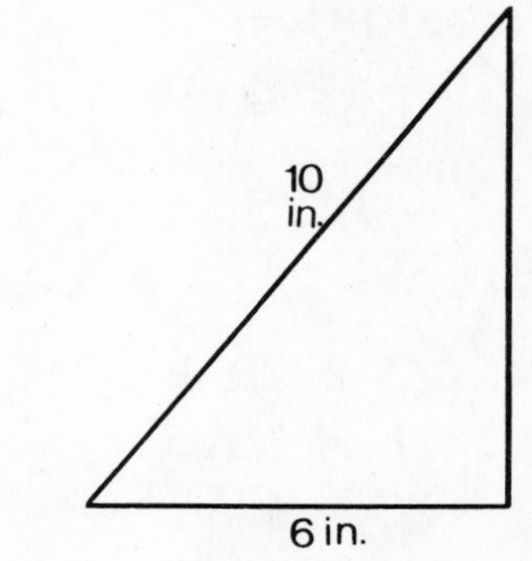

Step 1. Substitute 10 for c and 6 for a in the Pythagorean Theorem.

$c^2 = a^2 + b^2$
$10^2 = 6^2 + b^2$
$100 = 36 + b^2$

Step 2. Subtract 36 from each side, and solve for b.

$100 - 36 = b^2$
$64 = b^2$
$\sqrt{64} = b$
$8 = b$

Answer: **length of unlabeled side = 8 inches**

Solve the problems below.

1. The two sides of a right triangle measure 6 feet and 8 feet. What is the length of the hypotenuse?

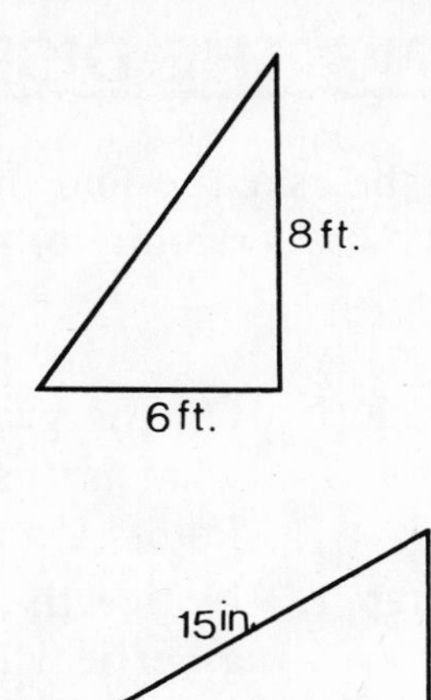

2. The hypotenuse of a right triangle is 15 inches long. If one side measures 12 inches, what is the length of the other side?

3. Bill had a garden in the shape of a right triangle. One side measured 5 yards and the other side measured 12 yards. How long was the third side —the side opposite the corner?

4. A ladder leaned against the side of the house. The length of the ladder is 17 feet. The base of the ladder is measured to be 8 feet from the house. How far off the ground is the top of the ladder?

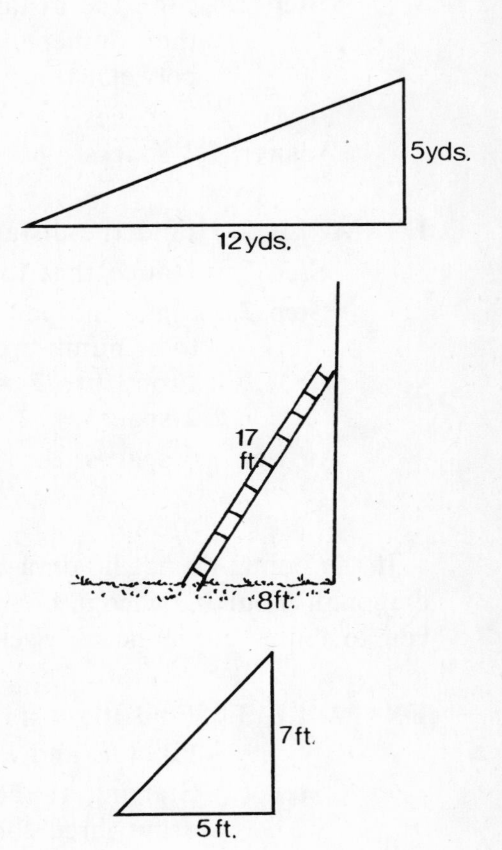

5. In a right triangle, one side measures 5 feet and the second side measures 7 feet. Find the approximate length of the hypotenuse.

6. A ship sails 10 miles west and 7 miles north of the harbor. Approximately how far is the ship from the harbor?

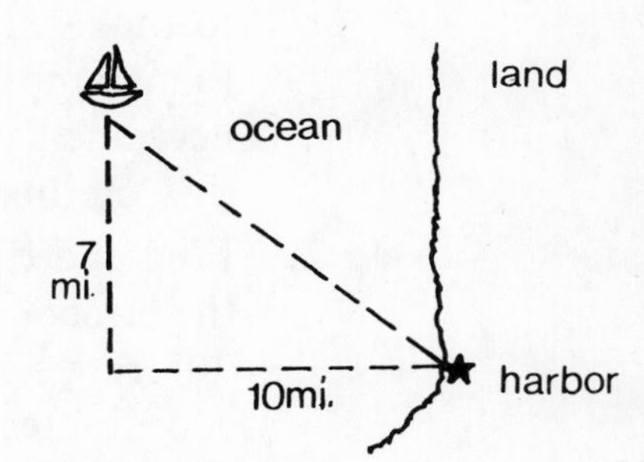

FINDING THE DISTANCE BETWEEN POINTS ON A GRAPH

You may be asked to find the distance between points on a graph. This is fairly simply if the points lie on a horizontal or vertical line. Simply find the number of spaces between the points.

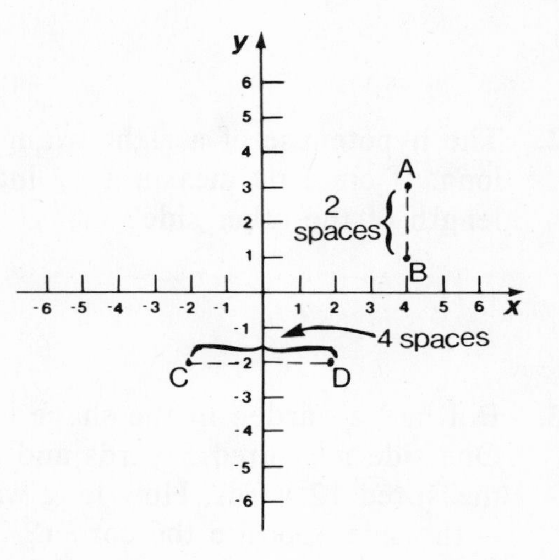

EXAMPLE 1. On the graph at the right, what is the distance between Point A and Point B?

Step 1. Notice that the points lie on a vertical line parallel (running in the same direction) to the y-axis.

Step 2. Find the distance by finding the number of spaces between the y coordinates.
3 spaces − 1 space = **2 spaces**

Answer: **2 spaces**

EXAMPLE 2. Find the distance between Point C and Point D.

Step 1. Notice that the points lie on a horizontal line parallel to the x-axis.

Step 2. Since the points are on opposite sides of the y-axis, you must add the total number of spaces on both sides. From − 2 to 0 = 2 spaces.
From 0 to 2 = 2 spaces
2 spaces + 2 spaces = **4 spaces**

Answer: **4 spaces**

If the points do not lie on a horizontal or vertical line, they lie on a slanted line called a diagonal. Your knowledge of the Pythagorean Theorem from the previous lesson will help you to find the distance between two points that lie on a diagonal.

EXAMPLE 3. Find the distance between Point E and Point F on the graph.

Step 1. Starting at the given points, draw three connecting lines to make a right triangle. (This step will not be necessary once you see how this method works.)

Step 2. Find the difference between the x coordinates and the difference between the y coordinates. Since each distance is always a positive number, subtract the smaller number from the larger number.

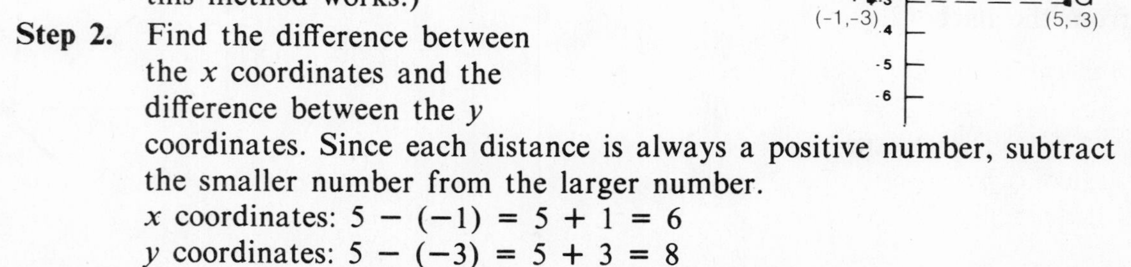

x coordinates: $5 - (-1) = 5 + 1 = 6$
y coordinates: $5 - (-3) = 5 + 3 = 8$

Step 3. Substitute these values into the Pythagorean Theorem and solve. (As you can see from the drawing, the diagonal from Point E to Point F forms the hypotenuse of a right triangle, and the numbers that you found by subtracting are the lengths of the other two sides of the triangle.)

$$c^2 = a^2 + b^2$$
$$(EF)^2 = 6^2 + 8^2$$
$$(EF)^2 = 36 + 64$$
$$EF = \sqrt{100}$$
$$EF = \mathbf{10\ spaces}$$

Answer: **10 spaces**

Find the distance between the points on the graph below.

1. Points A and B

2. Points A and C

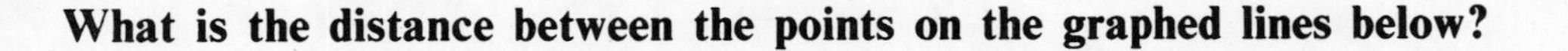

What is the distance between the points on the graphed lines below?

3. Points A and B

4. Points B and C

5. Points A and C

6. Points A and D. *Hint:* Before substituting values into the Pythagorean Theorem, rewrite the distances between the coordinates as improper fractions with a denominator of 2.

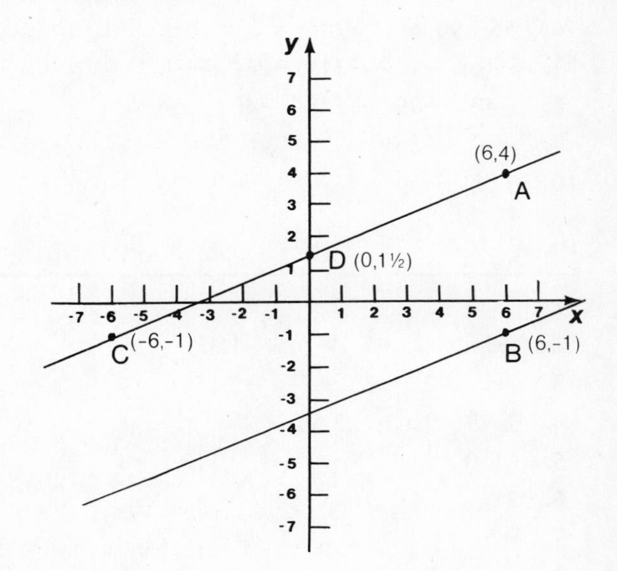

ANSWER KEY

PAGE 4

1. +8 +3 +12 +7
2. −5 −4 −19 −8
3. $+2\frac{1}{2}$ $+4\frac{3}{4}$ $-\frac{2}{3}$ $-\frac{1}{4}$
4. positive negative positive negative
5. positive positive negative positive

PAGE 5

1. -7.75 -7 -4 $-3\frac{1}{4}$ $+\frac{1}{2}$ +5 +6.5 +8

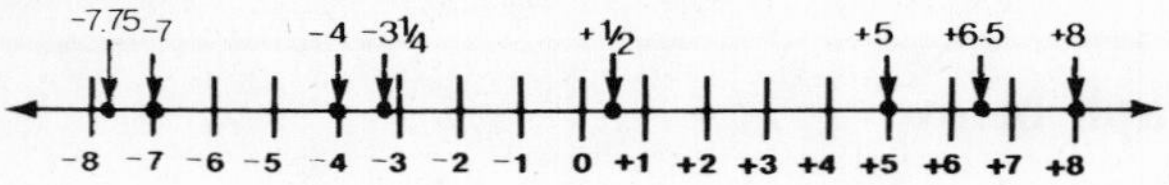

-8 -7 -6 -5 -4 -3 -2 -1 0 +1 +2 +3 +4 +5 +6 +7 +8

2. 0 +1 +2 +3 +4 +5 +6 +7

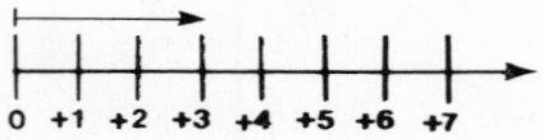

3. -7 -6 -5 -4 -3 -2 -1 0

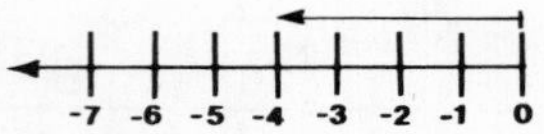

PAGE 7

1. 15 23 $33\frac{1}{2}$ 25 $19\frac{3}{4}$
2. 80 92 262 $798\frac{1}{2}$ 1,166.56
3. 36 82 540 437 561
4. 17 21
5. 159 114
6. 55 90
7. −16 −11 −11 $-26\frac{3}{5}$ −36.5
8. −43 −99 −148 −407 −614
9. −13 −35
10. −28 −54

PAGE 9

1. 2 5 2 $2\frac{7}{9}$ 2.5
2. −3 −3 −3 $-1\frac{1}{4}$ −1.3
3. −5 −2 −3 $-2\frac{1}{4}$ −2.15
4. 2 9 9 $3\frac{3}{4}$ 3.50
5. 0 0 0 0 0
6. 2 7
7. −5 −10
8. −2 −9
9. +17 +4
10. 0 0

PAGE 10

1. 1 −2 3 −11 $-3\frac{1}{7}$ 19
2. −6 −25 −11 9 0 −9.50
3. 0 −17 5 −6 0 13.80
4. 87 −130 35 44 $36\frac{2}{3}$ 12.41
5. 18 −19 −62 34 −58 −157
6. 127 166 −90 −200 228 −544
7. −15 −15 19
8. −12 −3 26
9. 30 −18 83
10. −66 120 −100

PAGE 11

1. −8
2. −12
3. −7
4. +16
5. −4
6. −7.50
7. $+2\frac{1}{2}$
8. $-4\frac{1}{4}$
9. −5
10. +5

PAGE 13

1. +2 +1 0 +5 +13
2. −2 −1 −3 −5 −12
3. −3 +5
4. +3 +6
5. −1 −9
6. +6 +11 −19 −4 +13 +3
7. +12 +30 −31 −20 +7 +6
8. +14 −4 −11 0 +14 −1
9. +1 +16 −30
10. +38 +12 −34

PAGES 14–15

1. +27 +42 +90 +132 +140
2. +15 +28 +72 +65 +168
3. −42 −24 −36 −77 −368
4. +15 +56 +45
5. +56 +48 +32
6. −10 −24 −49
7. −48 +45 +35 +18 −247 −432
8. −81 −8 −42 −392 +143 +1302
9. −130 +153 −864 −493 +1036 −3060
10. +84 −72 −52 −156
11. −255 +378 −24 +600

PAGE 17

1. $+4 \quad \frac{1}{3} \quad +4 \quad \frac{3}{5}$
2. $+5 \quad +3 \quad +\frac{3}{4} \quad +\frac{4}{5}$
3. $-\frac{1}{3} \quad -2 \quad -4 \quad -\frac{1}{3}$
4. $4 \quad +\frac{7}{10}$
5. $+7 \quad +\frac{3}{5}$
6. $-2 \quad -\frac{1}{9}$
7. $-4 \quad +4 \quad 1 \quad -7 \quad -\frac{3}{5}$
8. $-\frac{1}{3} \quad -5 \quad -3 \quad -\frac{2}{3} \quad -1\frac{7}{12}$
9. $-5 \quad -\frac{4}{7} \quad -5 \quad 3 \quad +8$
10. $-9 \quad +9 \quad +\frac{1}{3}$

PAGE 18

1. +21
2. +15
3. −10
4. +29

PAGE 19

1. +140
2. −162
3. −60
4. +168
5. −108
6. +192
7. +180
8. +120
9. −168

PAGE 21

1. 13°
2. 111°
3. 20,600 feet
4. 14,775 feet
5. $527.55

PAGES 22–23

1. a. +13 b. −26 c. $+2\frac{3}{4}$ d. $-\frac{7}{8}$

2. 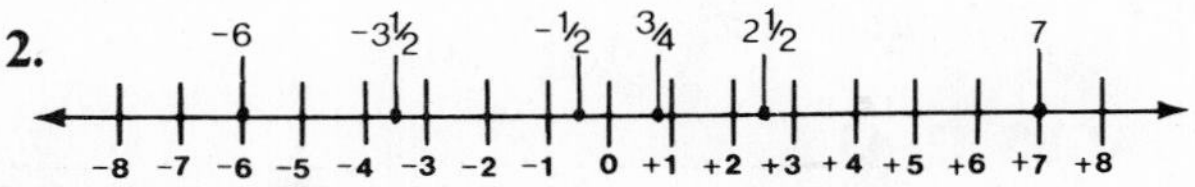 Number line from −8 to +8 with points marked at −6, $-3\frac{1}{2}$, $-\frac{1}{2}$, $\frac{3}{4}$, $2\frac{1}{2}$, 7

3. −29
4. −30
5. +6
6. −7
7. $-9\frac{1}{3}$
8. 0
9. +10
10. −25
11. +46
12. −26
13. 0
14. +136
15. +91
16. −116
17. −56
18. +3
19. $+\frac{3}{5}$
20. −4
21. −4
22. 23°
23. $303.75

PAGES 25–26

2. $(+5)^2$ positive five to the second power or positive five squared
3. $(+\frac{1}{2})^2$ positive one-half to the second power or positive one-half squared
4. 6^2 six to the second power or six squared
6. $(-3)^3$ negative three to the third power or negative three cubed
7. $(+6)^3$ positive six to the third power or positive six cubed
8. $(-\frac{2}{3})^3$ negative two-thirds to the third power or negative two-thirds cubed
10. $(5)^4$ five to the fourth power
11. $(-4)^4$ negative four to the fourth power
12. $(+\frac{1}{4})^4$ positive one-fourth to the fourth power

13. +125
14. +4
15. −216
16. +81
17. +625
18. $+\frac{4}{9}$
19. −1,000
20. −27
21. 625
22. $-\frac{64}{125}$
23. 25
24. −8
25. 55
26. 256
27. 72
28. 14
29. 56
30. $\frac{3}{8}$
31. −21
32. $\frac{13}{16}$

PAGE 27

1. 1,323
2. 1,125
3. −128
4. 5
5. $1\frac{11}{16}$
6. $\frac{1}{250}$
7. $1\frac{32}{49}$
8. −4
9. $\frac{4}{9}$
10. $2\frac{13}{18}$
11. $-\frac{25}{216}$
12. $\frac{1}{45}$

PAGE 28

2. $7 = \sqrt{49}$
3. $10 = \sqrt{100}$
4. $12 = \sqrt{144}$
5. $15 = \sqrt{225}$

ANSWER KEY (continued)

PAGE 29

1. 13 5 9
2. 2 11 6
3. 7 1 15
4. 12 8 3
5. 10 4 14
6. $\frac{2}{3}$ $\frac{3}{8}$ $\frac{5}{6}$
7. $\frac{1}{2}$ $\frac{7}{9}$ $\frac{11}{12}$
8. $\frac{9}{4}$ $\frac{1}{14}$ $\frac{10}{11}$

PAGE 31

Answers will vary if you start with a different value in Step 1.

1. Step 1: 6, Step 3: 6.5
2. Step 1: 4, Step 3: 4.375
3. Step 1: 8, Step 3: 8.75
4. Step 1: 9, Step 3: 9.5
5. Step 1: 5, Step 3: 5.3
6. Step 1: 10, Step 3: 10.35

PAGES 32–33

1. 3 base, 4 exponent
2. $(-4)^4$
3. $(+5)^3$
4. $(-\frac{3}{5})^2$
5. 8^2
6. $(+\frac{5}{6})^3$
7. 9^4
8. +81
9. −125
10. +81
11. $-\frac{8}{27}$
12. 256
13. $\frac{16}{25}$
14. 1
15. 7
16. 45
17. −16
18. 36
19. 7
20. $\frac{9}{11}$
21. 7.5
22. 9.11

PAGES 34–35

1. $x+9$
2. $2y$
3. $y+(-2)$
4. $5x$
5. $z-1$
6. $3x$
7. $\frac{r}{22}$
8. $x+(-4)$
9. $w-7$
10. $z-12$
11. $\frac{x}{5}$
12. $\frac{-x}{12}$
13. $c-4$
14. $y-(-9)$
15. $3+n$
16. $m-3$
17. $\frac{13}{y}$
18. $\frac{r}{-22}$
19. h
20. f
21. g
22. i
23. b
24. e
25. j
26. a
27. d
28. c

PAGES 36–38

1. 5
2. −9
3. 7
4. 12
5. 26
6. 8
7. 13
8. −18
9. 10
10. 70
11. 3
12. −3
13. −8
14. 22
15. 21
16. 27
17. 8
18. 7
19. 10
20. 9
21. −3
22. 16
23. 616
24. 5
25. −6
26. 25
27. 27
28. 10
29. 12
30. −20
31. $-\frac{1}{3}$

PAGES 39–40

1. P = 22 feet
2. P = 28 feet
3. C = 44 inches
4. A = 84 square feet
5. A = 42 square inches
6. A = 154 square inches
7. V = 630 cubic inches
8. V = 88 cubic inches
9. V = $113\frac{1}{7}$ cubic inches

PAGE 41

1. 18 feet
2. 1399 feet
3. 176 feet
4. 180 square feet
5. 180 square feet
6. 154 square meters
7. 756 cubic feet

PAGES 42–43

1. 12 plus y
2. x minus 9
3. 15 times a
4. 13 divided by y
5. negative 9 times the quantity "2 times x plus 4"
6. $x - 7$
7. $\frac{z}{9}$
8. $z + 19$
9. $8m$
10. $-4(y - 8)$
11. −2
12. 34
13. +4

PAGE 43 continued

14. 44
15. −6
16. 32
17. 7
18. $+16\frac{2}{3}$
19. $282\frac{6}{7}$ cubic feet
20. $99
21. 180 square feet
22. 40° C

PAGE 44

More than one expression is possible for each equation.

1. x plus 52 equals 72
2. y minus 21 equals 13
3. 7 times a equals 147
4. z divided by 6 equals 16

PAGE 45

1. No
2. Yes
3. Yes
4. No
5. Yes

PAGE 47

1. 4
2. 2
3. 14
4. 11
5. −4
6. −7

PAGE 48

1. 12
2. 11
3. 55
4. 65
5. 6
6. −9

PAGE 49

1. 7
2. $\frac{1}{4}$
3. −4
4. $-\frac{2}{3}$
5. −9
6. 17

PAGE 50

1. 35
2. 72
3. 4
4. 2
5. −56
6. −63

PAGE 51

1. 12
2. 49
3. −24
4. 18

PAGES 52–53

1. 2
2. 4
3. −9
4. 3
5. 19
6. 25
7. −4
8. 15
9. 8
10. −8
11. 11
12. −9
13. 27
14. 56
15. 28
16. −24
17. 15
18. −21
19. 36
20. −8
21. 12
22. −24
23. 9
24. 7
25. 20
26. −3
27. 11
28. 32
29. 60
30. −4
31. 9
32. −24
33. 1
34. −2
35. 15
36. 5
37. 9
38. −6
39. −9
40. −12
41. $-\frac{19}{5}$ *or* $-3\frac{4}{5}$
42. −2
43. −9
44. 45

PAGE 55

1. 11
2. 21
3. 7
4. 1,452
5. 540
6. $770
7. $95
8. $9,475

PAGE 57

1. 3
2. −4
3. 8
4. −7
5. 2
6. −30
7. 4
8. −2

PAGE 59

1. 7
2. 5
3. 5
4. 6
5. 18
6. 70
7. 1
8. 3
9. 3
10. 5
11. 9
12. 1
13. −1
14. 4
15. 21

PAGES 60–61

1. 4
2. 4
3. 8
4. 5
5. 2
6. 26
7. 35
8. 18
9. 4
10. −7½
11. 2
12. 10
13. 32
14. 5
15. 3
16. 2
17. 6

ANSWER KEY (continued)

PAGES 62–63

1. 3	**6.** 2	**11.** 2
2. 9	**7.** 4	**12.** 8
3. 3	**8.** 3	**13.** 3
4. −3	**9.** −9	**14.** −23
5. 8	**10.** −10	**15.** $\frac{3}{2}$

PAGES 64–65

1. 8
2. 13
3. 24
4. 30
5. $75 Terry's share
   $150 Susan's share
6. $300 Manuel's share
   $450 Jorge's share
7. 27 appliances
8. $150

PAGE 66–69

1. 5	**10.** 4	**19.** −12
2. 4	**11.** 8	**20.** 4
3. 4	**12.** 15	**21.** −6
4. −4	**13.** 4	**22.** 3
5. 21	**14.** 4	**23.** 4
6. 20	**15.** 6	**24.** −3
7. 1	**16.** 2	**25.** 3
8. 3	**17.** 3	**26.** −15
9. 16	**18.** 2	

PAGES 70–71

1. 5
2. 16
3. 11
4. 6
5. $85 Maria's expenses
   $75 Amy's expenses
   $150 Sadie's expenses
6. $2.65 Frank's meal
   $2.20 Sam's meal
   $4.40 Louis's meal
7. 17 18 19
8. 20 22 24

PAGE 72

1. $\frac{3}{5}$ *or* 3:5	**6.** $\frac{5}{3}$ *or* 5:3
2. $\frac{3}{4}$ *or* 3:4	**7.** $\frac{3}{1}$ *or* 3:1
3. $\frac{25}{7}$ *or* 25:7	**8.** $\frac{5}{1}$ *or* 5:1
4. $\frac{1}{3}$ *or* 1:3	**9.** $\frac{2}{1}$ *or* 2:1
5. $\frac{3}{2}$ *or* 3:2	

PAGE 73

1. $\frac{3}{2}$ *or* 3:2	**4.** $\frac{1}{3}$ *or* 1:3
2. $\frac{2}{3}$ *or* 2:3	**5.** $\frac{1}{2}$ *or* 1:2
3. $\frac{6}{7}$ *or* 6:7	

PAGE 75

1. $\frac{6}{12} = \frac{9}{18}$	**12.** $\frac{6}{4} = \frac{12}{z}$
2. $\frac{14}{20} = \frac{7}{10}$	**13.** $\frac{30}{24} = \frac{x}{12}$
3. $\frac{2}{5} = \frac{30}{75}$	**14.** 3
4. $\frac{10}{2} = \frac{5}{1}$	**15.** 6
5. $\frac{9}{3} = \frac{12}{4}$	**16.** 5
6. $\frac{5}{10} = \frac{20}{40}$	**17.** 20
7. $\frac{6}{3} = \frac{16}{8}$	**18.** 10
8. $\frac{3}{4} = \frac{15}{20}$	**19.** 3
9. $\frac{25}{10} = \frac{5}{2}$	**20.** 60
10. $\frac{x}{7} = \frac{3}{21}$	**21.** 27
11. $\frac{5}{y} = \frac{15}{20}$	**22.** 40

PAGE 77

1. 5 inches
2. $1.35
3. 4 hours
4. $88.50
5. 4 gallons of white paint
6. 12 drops of hardener
7. 612 miles
8. $2.61
9. 6 quarts of brown paint
10. 5.08 centimeters

PAGE 79

1.

y
4
5
6
7

3.

a
5
3
1
−1

5.

y
2
4
6
8

2.

y
1
4
7
10

4.

m
2
$2\frac{1}{2}$
3
$3\frac{1}{2}$

6.

y
6
3
0
−3

PAGES 80–83

1. 23
2. 25
3. 18
4. 1,107
5. 33
6. 8
7. $11,250
8. 4
9. 9
10. 2
11. 36
12. 4
13. 13
14. $91 Mary's share
$65 Lucy's share
15. 8
16. 4
17. 9
18. 24 hours—Alice
16 hours—Joan
32 hours—Judy
19. $\frac{9}{13}$ *or* 9:13
20. $\frac{2}{1}$ *or* 2:1
21. 6
22. $45.50
23. $y = 4(x+9)$
24.

y
1
5
9
13

PAGE 85

1. A: x coordinate = 5
y coordinate = 2
B: x coordinate = −3
y coordinate = 4
C: x value = −2
y value = −3
D: $x = 4$
$y = -3$

PAGES 86–87

1.	(+3,+5)	(−4,−2)	(7,0)	(0,−3)
2.	$(+\frac{2}{3},-7)$	(6,1)	$(-3,\frac{1}{2})$	$(-\frac{3}{4},-2\frac{1}{4})$
3.	$x = +2$	$x = -3$	$x = -7$	$x = \frac{1}{2}$
	$y = +6$	$y = +4$	$y = -5$	$y = +5$
4.	$x = -3$	$x = \frac{2}{3}$	$x = 4$	$x = -\frac{2}{3}$
	$y = 2$	$y = -4$	$y = 7$	$y = -1$

5. A = (5,4)
B = (2,3)
C = (−5,1)
D = (−3,−4)
E = (4,−2)
F = (1,−5)

6. A = $(\frac{1}{2},5)$
B = (0,0)
C = $(-2\frac{1}{2},1\frac{1}{2})$
D = (−6,−3)
E = (1,−2)
F = (4,−5)

7. A = (6,3)
B = $(3,1\frac{1}{2})$
C = (2,1)
D = (−2,−1)
E = (−4,−2)
F = (−6,−3)

8. A = (6,−2)
B = (4,−1)
C = (2,0)
D = (0,1)
E = (−2,2)
F = (−4,3)

PAGES 88–89

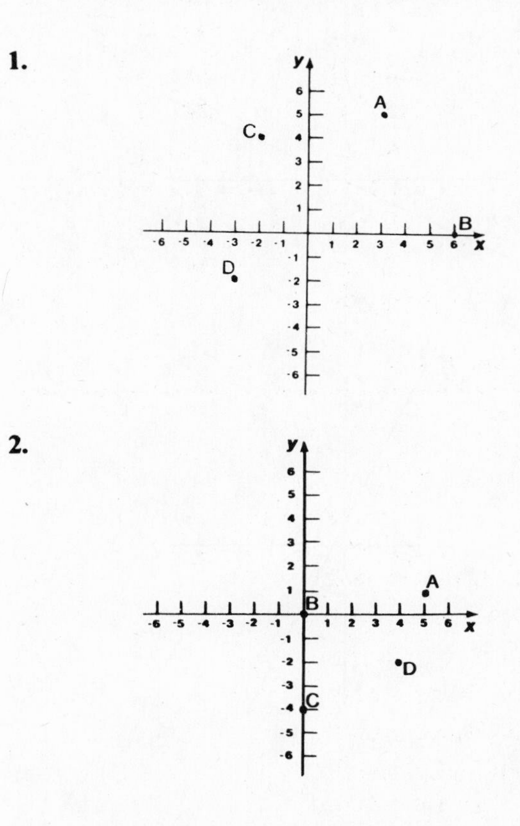

ANSWER KEY (continued)

Pages 88–89 continued

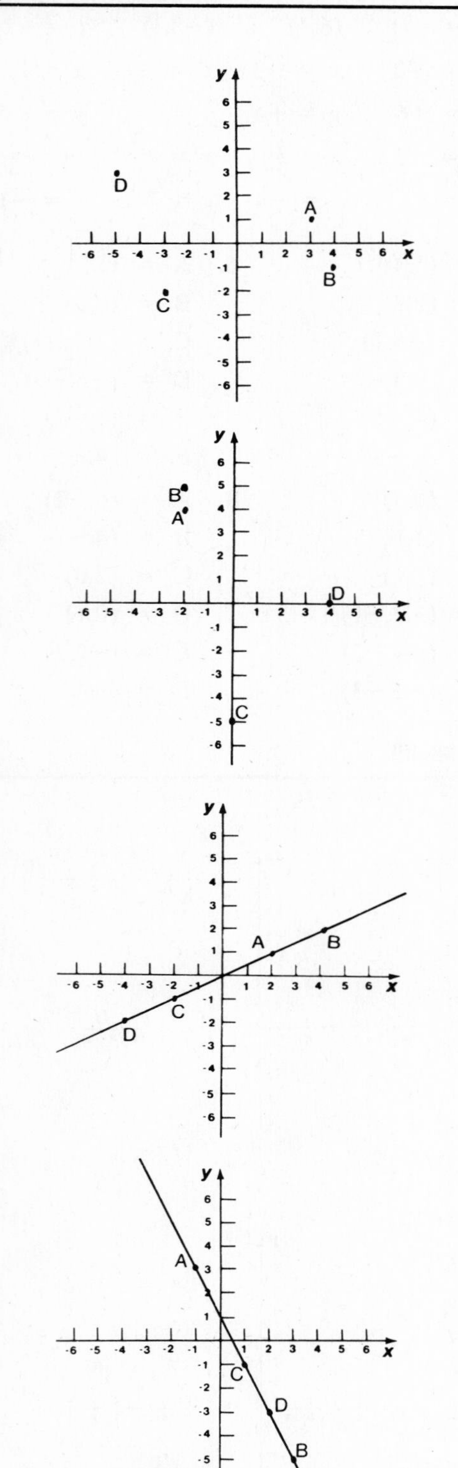

PAGES 90–91

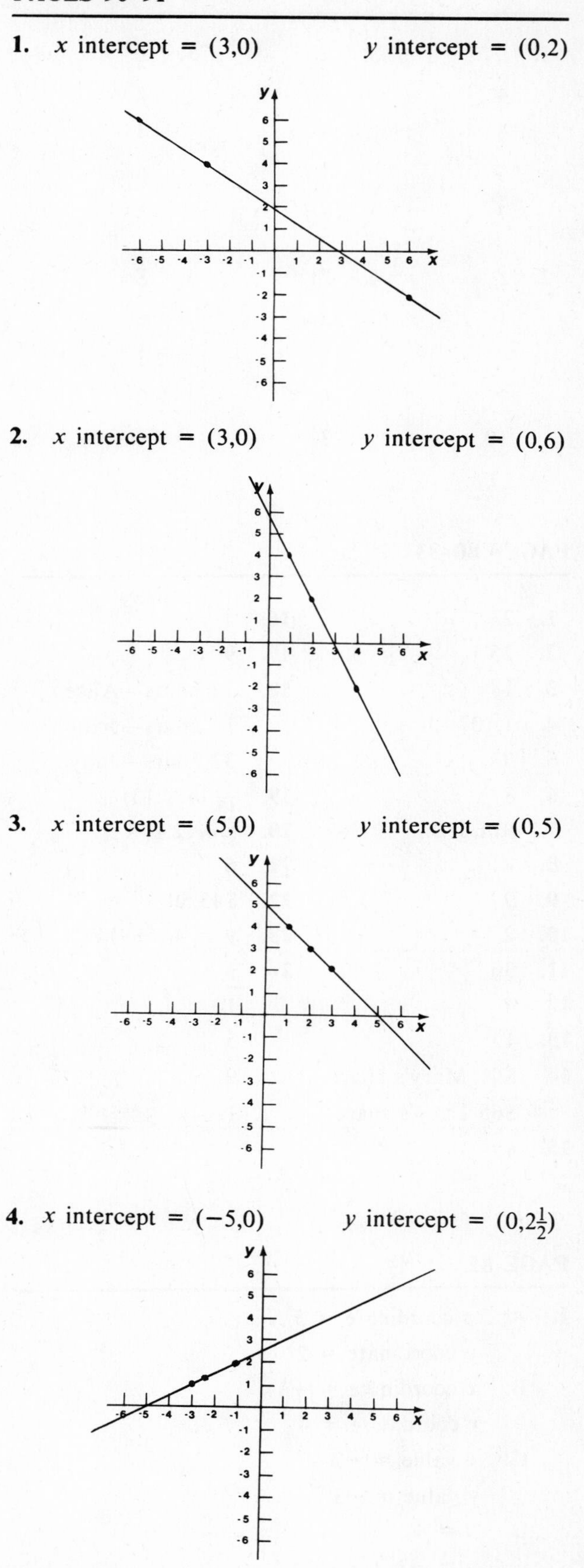

1. x intercept = (3,0) y intercept = (0,2)

2. x intercept = (3,0) y intercept = (0,6)

3. x intercept = (5,0) y intercept = (0,5)

4. x intercept = (−5,0) y intercept = $(0,2\frac{1}{2})$

5.

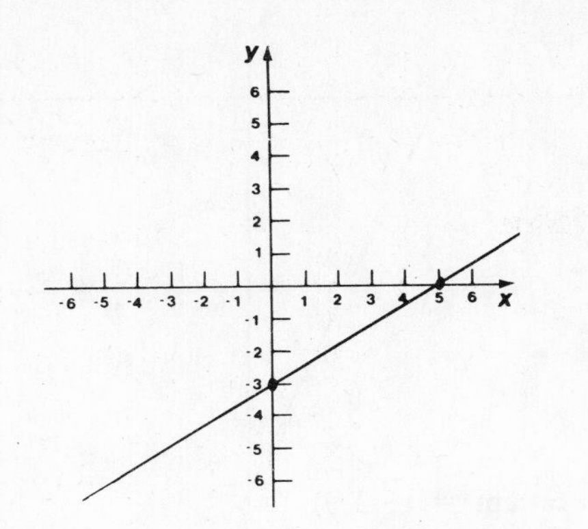

6.

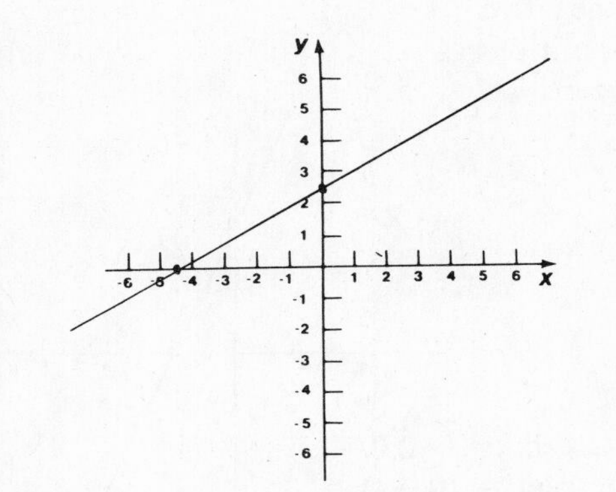

PAGE 92

1. A. negative
B. positive
C. zero

2. D. negative
E. positive
F. undefined

PAGE 93

1. 1
2. $\frac{2}{3}$
3. $-\frac{1}{2}$
4. -2
5. $-\frac{2}{3}$

PAGES 94–96

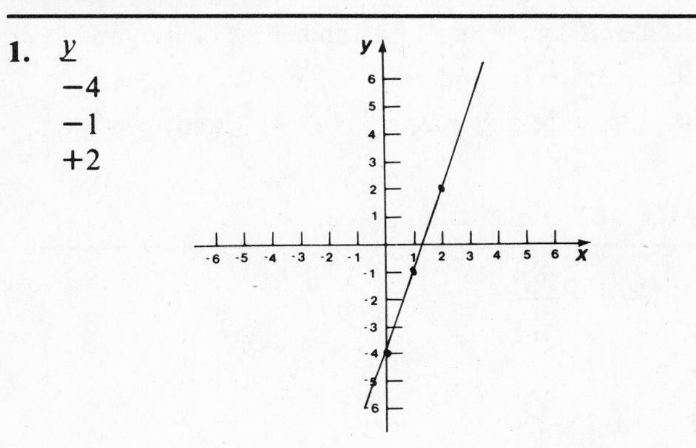

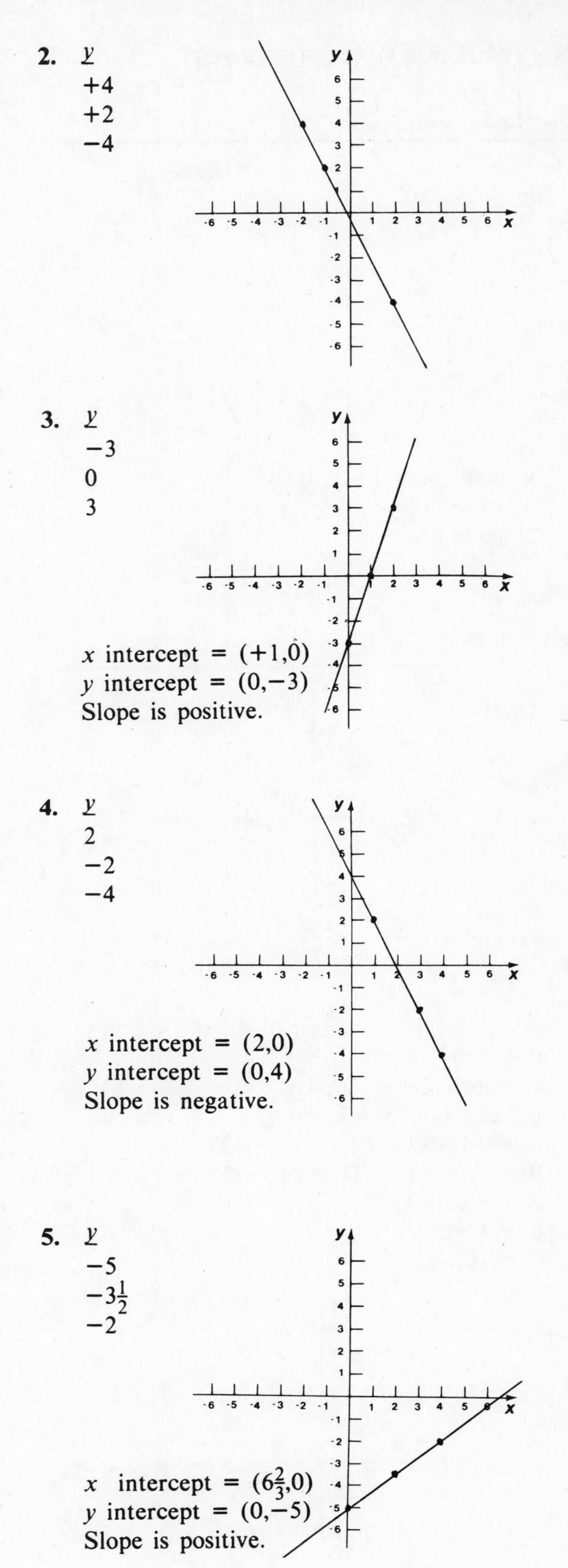

ANSWER KEY (continued)

Pages 94–96 continued

6. $\underline{y}$
-1
1
3

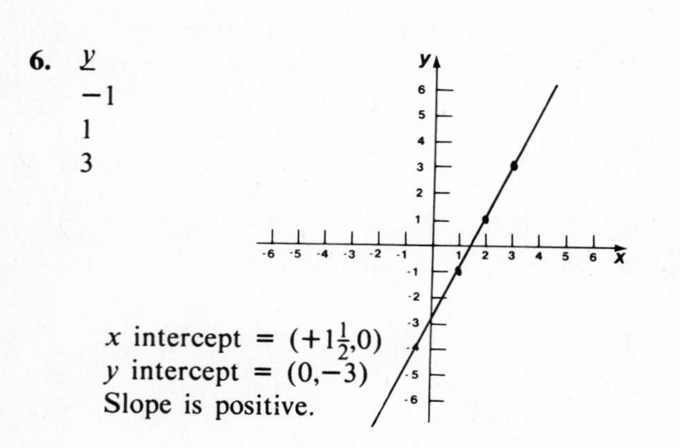

x intercept = $(+1\frac{1}{2},0)$
y intercept = $(0,-3)$
Slope is positive.

PAGE 97–99

1.

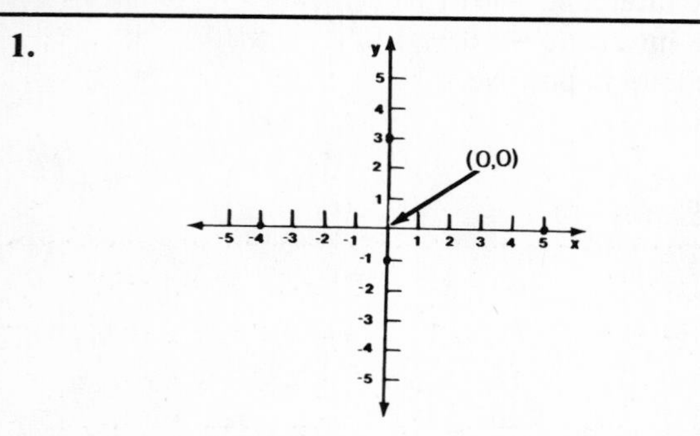

2. $(-6,-3)$

3. x coordinate = -4
y coordinate = $+5$

4. A = $(+6,+1)$ C = $(-4,-2)$
B = $(-5,+3)$ D = $(+3,-3)$

5. A = $(-5,+3)$
B = $(-3,0)$
C = $(0,-5)$

6.

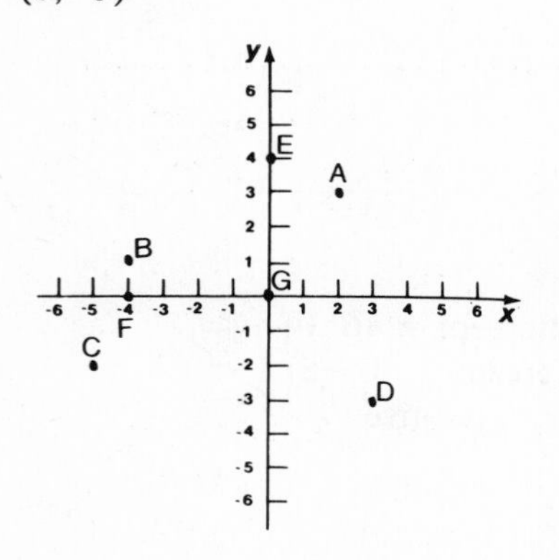

7.

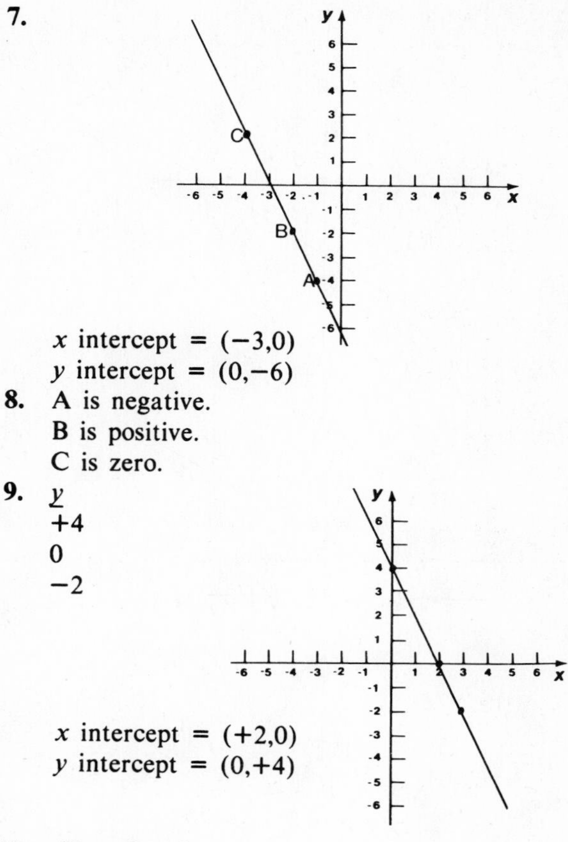

x intercept = $(-3,0)$
y intercept = $(0,-6)$

8. A is negative.
B is positive.
C is zero.

9. $\underline{y}$
$+4$
0
-2

x intercept = $(+2,0)$
y intercept = $(0,+4)$

10. Slope is -2.

PAGE 100

1. binomial
2. trinomial
3. monomial
4. binomial
5. trinomial
6. monomial
7. $-9d$ and $+6$ $\quad 3a^2$ and $-5a$ $\quad -4y$ and $-\frac{1}{2}$
8. $-6d$ and $+8e$ $\quad 8x^2$ and $+3y^2$ $\quad \frac{2}{3}z^2$ and $-7z$
9. $-3y^2, -7y$, and -8 $\quad 8x, +4y$, and -13
10. $2a, +3b$, and $-2c$ $\quad 4u^2, -v^2$, and $+\frac{3}{4}w^2$

PAGE 101

	C.	V.P.
1.	$+3$	a
2.	-5	z
3.	$+1$	z^2
4.	-1	a^2
5.	-4	c^2d

6. $7 \quad uv^2$
7. $8 \quad x^2y^2$
8. $-4 \quad ab^3$
9. x and $-5x$
10. a^2 and $-4a^2$
11. $+2xy^2$ and $+xy^2$
12. -2 and $+4$
13. x^3 and $4x^3$
14. $-rs^2$ and $+3rs^2$
15. y^3 and $2y^3$
16. $+c^2$ and $-5c^2$
17. x^2y^2 and $-x^2y^2$

PAGES 102–103

1. $6y \quad 3z \quad 6n \quad 7r \quad 7c$
2. $5z \quad -3c \quad -2u \quad r \quad x$
3. $-10z \quad -13y \quad -15n \quad -7y \quad -9a$
4. $-4ab^2 \quad -3cd \quad -4xyz \quad -11a^2b \quad 5x^2y^2$
5. $2a^2b \quad 5x^2y \quad -c^2d \quad 5xy \quad 20z$
6. $2c^2-3d \quad -4ab^2 + 2ab \quad 2yz + xy$
   $7r^2 - s^2 \quad -6t - 4u$
7. $21yz \quad 20xy^2$
8. $-b^2 \quad 7c$
9. $11xy \quad -13d^2$
10. $-10x \quad -x^2 \quad -7y + 3z \quad 4r^2 \quad -7xy \quad 5uv$
11. $7c -5d \quad -8b^2 \quad 7a^2b \quad 3u^2 \quad -5x^2y \quad -5v$
12. $3xy^2 \quad 3b \quad 2ab + 3ad \quad 20a \quad -13v \quad 9r$
13. $15a^2 \quad 2xy \quad -17cd \quad 6r^2 \quad -11xy^2 \quad -ab$
14. $3x^2 \quad -2ab \quad 18xyz$
15. $-15a^2b \quad 9x \quad 8yz$
16. $0 \quad 14x^2y^2 \quad -8y$

PAGES 104–105

1. $3y - 4 \quad 5z - 2 \quad 3a + 5 \quad 7u - 9$
2. $4x - 5 \quad 6a + 8 \quad b + 3 \quad -2x - 7$
3. $4y + 6 \quad 2z + 10 \quad 7c + 7 \quad 7x + 11$
4. $3z - 3 \quad 10c + 1 \quad 5a \quad 4u - 2$
5. $9y^2 - 3y \quad 3a^2 + 6a \quad 2u^2 + 6v \quad 5r^2$
6. $5a - 2b + 4c \quad r + 5s - t \quad 8x - 2y + z$
7. $-2a + 8b \quad 6r \quad 10x + 2y \quad 5a - 10b$
8. $7x + y + 3z \quad 8x + y - z \quad x^2 + 2x - 1$
   $3y^2 + 2y + 2$

PAGES 106–107

1. $-3z \quad a \quad 5c \quad -3a \quad -4x$
2. $14a \quad -10n \quad 16r \quad -15z \quad -13y$
3. $c \quad -5y \quad +2d \quad -3s \quad 0$
4. $4a^2b^2 \quad -13xyz \quad n^2 \quad 0 \quad 2x^2y$
5. $3x^2 - x^3 \quad -5cd - 8c^2 \quad 6r^2s + 2rs$
   $-9u^2v - uv^2 \quad -12x^2 + 8x$
6. $6y \quad 3a^2b$
7. $13b \quad 19x$
8. $-5x \quad +3z$
9. $2y \quad 4z$
10. $17x \quad 3z$
11. $3z \quad 5x^2 - 7xy \quad -17c^2d \quad 10xy$
    $3uv \quad y$
12. $x^2y^2 + x^2y \quad 15y \quad 3c \quad 21r^2 \quad y^2 \quad -x$
13. $-y \quad -14x \quad 4z \quad 17x^2 \quad +3y^2 \quad -34z$
14. $13x^3 + 9x^2 \quad c^2d \quad 8rs + rs^2 \quad -17t^2$
    $-9x \quad 7u$
15. $-3x \quad 5y \quad 30x^2$
16. $-x \quad y^2 \quad 8t$

PAGES 108–109

1. $2z -1 \quad 7n - 3 \quad 2s + 2 \quad 4x - 3$
2. $5y - 5 \quad 7x + 7 \quad 9y - 8 \quad z + 2$
3. $2x - 2 \quad 3y - 13 \quad z + 5 \quad 7x + 8$
4. $3a + 4b \quad 7r + 8s \quad 7y + 7z \quad 10u - 8$
5. $y^2 - 6y \quad 3r^2 + r \quad 7u^2 + 7u \quad 5x^2 + 2$
6. $2a^2 - 7a - 4 \quad 5r^2 + r + 1 \quad x^2 + 8x - 8$
7. $5a - 7b \quad 8r - s \quad u \quad 6x - 4y$
8. $2x + y + 4z \quad -2y \quad x^2 -4x + 9$
   $-4a^2 + 9a - 8$
9. $10r \quad 3x - 2y + 8z \quad 4x^2 + 8$
   $3a - b + 5c$

PAGE 110

1. $8y \quad -21z \quad -18a \quad -35c \quad -32b$
2. $-14xy \quad -12ab^2 \quad 32rs \quad 27y^2 \quad -60z^3$
3. $-16x^2y \quad 30a^2b^2$
4. $-2xyz \quad 3y^2z$
5. $-15a \quad -45x^2 \quad -35x^2y^3 \quad 18x \quad 12xy^2$
   $-24a$
6. $-48rs \quad 27abc \quad -30r^2 \quad 18n \quad 35r^2s^2 \quad 48b$
7. $-xy \quad -20xy^2 \quad -20x^2y^2$
8. $42ab^2 \quad -5c^2d \quad 52xyz$

ANSWER KEY (continued)

PAGE 111

1. y^2 z^2 a^2 c^2 d^2
2. x^4 a^3 z^4 d^5 c^3
3. y^2 z^2 a^2
4. x^4 a^3 b^4

PAGES 112–113

1. $6y^2$ $-2z^2$ $-10y^2$ $-28c^2$ $-18x^2$
2. $6xy$ $18ab$ $-8mn$ $8xz$ $-8br$
3. $6a^2b$ $6c^2d$ $-28x^2z$ $8m^2n$ $3yz^2$
4. $-8ac^2$ $-40y^2z$ $16r^2s$ $-6x^2y$ $18y^2z$
5. $12xy^3z$ $12x^3yz$ $-2a^2b^2c$ $-10r^2s^2t$ $-16a^3bd$
6. $12a^3b^3$ $8r^3s^2t^3$ $18r^5s^2t^3$ $24x^2y^3z^3$ $-2a^4b^5$
7. $-12a^2b^3$ $-10r^3s^2$
8. $6r^5s^2t$ $-8xy^2z^4$
9. $6a^2$ $15ab$ $-12x^2y$ $-15r^3s$ $21x^2$ $4a^3b^3$
10. $12a^2b$ $42x^2y^2$ $20a^4b^3$ $24y^2$ $-39rs^2t$ $18xy$
11. $-12yz$ $-15a^3bc$ $-14z^2$ $36a^2b^3$ $16r^3s^3$ $-12b^2$
12. $14x^3y^2$ $28ab$ $-12u^3v^4$ $27xz$ $-8x^3yz$ $24a^2bc$
13. $-6x^3y$ $4a^3b^5$ $6x^2y^2z^3$
14. $-6a^3b^3c$ $-8x^3y^3$ $-12x^3y^2z$

PAGES 114–115

1. $30a + 15$ $21y - 42$ $-32z - 20$ $96x - 32$
2. $12x^2 - 6x$ $56r^2 + 16r$ $-28z^2 + 7z$ $20a^2 - 8a$
3. $-20a^3 - 20a^2$ $-26z^3 + 2z^2$ $14b^3 + 8b^2$ $-24r^3 - 18r^2$
4. $15y^2 + 21y - 18$ $20z^2 - 25z - 20$ $-18a^2 + 48a + 30$
5. $-24x^3 + 36x^2 - 4x$ $36y^3 - 12y^2 + 15y$ $-4z^3 + 10z^2 - 6z$
6. $-30z + 20$ $-12x^2 + 20$ $35a^2 + 21a$
7. $-6z^2 - 3z + 21$ $-20n^2 - 5n + 15$ $27z^2 + 18z + 36$
8. $18n^3 - 48n$ $27x^3 - 36x^2$ $-8y^3 - 10y^2$
9. $6a^3 + 72a^2 - 30a$ $-6x^3 - 8x^2 + 6x$ $-6s^3 - 24s^2 + 12s$
10. $-15a^3 - 12a^2$ $-20x^2 - 30x + 15$ $18y + 6$ $-20c^2 + 60$
11. $8x^3 - 10x^2 + 6x$ $-42y + 21$ $7z^3 - 2z^2$ $-28a^3 - 12a^2$
12. $28b^3 + 12b^2$ $10y^3 + 6y^2 - 8y$ $24t^2 - 30t + 18$ $42s^2 + 56$

PAGES 116–117

1. $2y^2 + 13y + 15$ $6z^2 + 11z + 4$ $12a^2 - 10a - 12$
2. $6x^2 + 5xy - 4y^2$ $7c^2 - 19cd - 6d^2$ $8r^2 + 10rs - 12s^2$
3. $6a^2 - 13ab + 6b^2$ $y^2 - 5yz + 6z^2$ $10a^2 - 19ab + 6b^2$
4. $8y^2 + 16y + 6$ $8a^2 - 2$ $15z^2 + z - 6$
5. $4x^2 - xy - 3y^2$ $2a^2 - ab - 6b^2$ $3r^2 + rs - 2s^2$
6. $a^2 - 2ab + b^2$ $3c^2 + 5cd + 2d^2$ $2y^2 - 5yz + 2z^2$

PAGE 119

1. 1 1 1 1 1
2. z^2 y^3 a b^2 r^2
3. x^2 c b^2 c^2 x^2
4. $\frac{1}{a^3}$ $\frac{1}{y}$ $\frac{1}{x^4}$ $\frac{1}{z}$ $\frac{1}{b^2}$
5. $\frac{1}{x^2}$ $\frac{1}{a}$ $\frac{1}{b^2}$ $\frac{1}{c^3}$ $\frac{1}{d^2}$
6. $\frac{1}{x^2}$ $\frac{1}{a^2}$ t c 1 $\frac{1}{z^2}$
7. 1 y^2 1 $\frac{1}{x^2}$ c $\frac{1}{s^3}$
8. $\frac{1}{z}$ $\frac{1}{x}$ y^2 $\frac{1}{r}$ 1 $\frac{1}{t^3}$
9. $\frac{1}{a^3}$ c^2 $\frac{1}{x}$ 1 $\frac{1}{u}$ v^2

PAGES 120–121

1. $-3a$ $4c^2$ $-5y$
2. $-4uv$ $4e$ $-7y^3z$
3. $-4x$ $9c^2$ $2x$
4. $-\frac{3b^2}{2c}$ $3rs$ $\frac{-3xy^3}{2z}$
5. $\frac{5}{a^3}$ $\frac{-3}{c^2}$ $\frac{-3}{d}$
6. $\frac{-1}{2x^2}$ $\frac{3s}{r^3}$ $\frac{4}{u^2v^2}$
7. $2b$ $3b^2$ $\frac{u^2v^2}{3}$ $\frac{3a^2}{4}$
8. $\frac{2x^3}{z}$ $\frac{-2a^3}{b}$ $\frac{2}{3y^3}$ $13b^3$
9. $\frac{-7}{2z}$ $\frac{2x^2}{3y}$ $\frac{6cd^2}{7e}$ $\frac{3ab^3}{4}$
10. $\frac{3}{r^2}$ $-4z^3$ $\frac{u^2v}{3}$ $\frac{y^4}{3z}$
11. $3y^4$ $\frac{2y^3}{x^3}$ $\frac{-3a}{b}$ $\frac{-3}{2c^2}$
12. $-7s^2$ $3y^2$ $\frac{3y^2}{4}$ $\frac{b}{3a^2c^2}$

PAGES 122–123

1. $3y + 4$ $2z - \frac{1}{2}$
2. $\frac{7}{2}x^2 + \frac{7}{4}x$ $\frac{3}{2}z^2 - z$
3. $4a - 3$ $c + 2$
4. $\frac{1}{d^2} + \frac{4}{d^3}$ $\frac{2}{3u} - \frac{1}{3u^2}$
5. $2y^2 + 3y - 2$ $-2z^4 + z^2 - 1$
   $\frac{2a^2}{3} - \frac{a}{2} + \frac{1}{3}$
6. $6y^2z^2 + 3yz + 1$ $x^2y^2 - 2xy - 1$
   $u^3v^3 - 2uv - 3$
7. $5y - 3$ $3c - 1$ $-2x^2 - 1$
8. $2x - \frac{1}{2x}$ $-a - 2$ $7a^2 - 3a$
9. $5y - 2$ $\frac{z^3}{3} + \frac{z}{4}$ $\frac{9u^3}{8v} - \frac{v^2}{u}$
10. $\frac{3}{4c} + \frac{1}{c^2}$ $\frac{3d^2}{2} + \frac{1}{2}$ $\frac{2}{x^2} - \frac{1}{2x^3}$
11. $x^2 - 2x + 3$ $z^3 + 2z + 3$
    $-4u^4 - 2u^2 + u$ $\frac{y^2}{2} - \frac{y}{2} + 2$
12. $-a^2b^2 + 3ab - 2$ $2c^3d^3 - cd + 3$
    $x^2y^3 + \frac{3}{2}xy - 2$ $y^3 - 2y^2 + 3y$

PAGES 124-125

1. $5y$
2. $-13x^2$
3. $9c - 5d$
4. $11y^2 + y + 3$
5. $-3a + 20b$
6. $4y^2 - 2y - 9$
7. $21z$
8. $21a^2$
9. $7x - 11y$
10. $3c^2 - 3c + 9$
11. $11x - 15y$
12. $5d^2 + 2d + 2$
13. $15y^2$
14. z^9
15. $-10a^3b^3$
16. $10x^3 - 6x^2$
17. $12a^2 - 30ab + 12b^2$
18. $-6y^3 + 12y^2 - 33y$
19. $-10x^2 - 36x - 18$
20. $4z^4$
21. c^3
22. $\frac{1}{d^3}$
23. $7d$
24. $-4y^2 + 3y + 5$
25. $4a^2b^2 - 2ab + 3$

PAGES 126–130

1. +2
2. −13
3. −20
4. −7
5. −91
6. −4
7. 24° change
8. −27
9. $\frac{2}{27}$
10. $\frac{5}{8}$
11. 7
12. $\frac{8}{9}$
13. 6.5
14. $3(x+7)$
15. −12
16. 27 square feet
17. 32
18. −10
19. \$80 Joyce's earnings
    \$60 Vicki's earnings
20. −1
21. \$685 Bill's earnings
    \$645 John's earnings
    \$1,290 Jeff's earnings
22. 28
23. $\frac{1}{4}$ *or* 1:4
24. 24
25. \$540
26. $\frac{1}{4}$ mile
27. y
    5
    3
    1
    Slope is negative.
28. x intercept = (2, 0)
    y intercept = (0, 4)
29. $x^2 + x$
30. $-10y - 4z$
31. $-2x^4y^3$
32. $18x^2 + 12x$
33. $6a^2 - 4ab - 2b^2$
34. $-9ab$
35. $-3x^2 + 4x - 2$

PAGES 132–133

1. 18° difference
2. 35° difference
3. 40° difference
4. 49° difference
5. 6° difference
6. 47° difference
7. −10°F at 15 mph
8. −15°F at 15 mph

PAGE 135

1. 600 miles
2. 55 mph
3. 7 hours
4. 288 miles
5. 52 mph
6. 6 hours

PAGE 137

1. \$30
2. \$721.50
3. \$95
4. \$1,420
5. \$71.88
6. \$453.75

PAGE 139

1. \$67.50
2. \$1,320
3. \$85
4. \$1,112.50
5. \$112.50
6. \$70

ANSWER KEY (continued)

PAGE 141

1. $2,122.42
2. $1,125.51
3. $551.91

PAGE 143

1. $1\frac{5}{7}$ hours
2. $2\frac{2}{9}$ hours
3. $1\frac{1}{3}$ hours
4. $1\frac{7}{8}$ hours
5. $\frac{3}{5}$ hour

PAGE 145

1. 10 pounds
2. 4 pounds
3. 1 pound

PAGE 147

1. 8 gallons
2. 400 pounds
3. $1\frac{1}{3}$ quart

PAGES 148–149

1. Fahrenheit
2. Celsius
3. 0–100 degrees Fahrenheit
4. 0–40 degrees Celsius
5. 0° C
6. 15° C
7. 5° C
8. 20° C
9. 35° C
10. 30° C
11. 32° F
12. positive
13. 20°
14. 210°
15. 40°
16. 100°

PAGE 151

1–5.

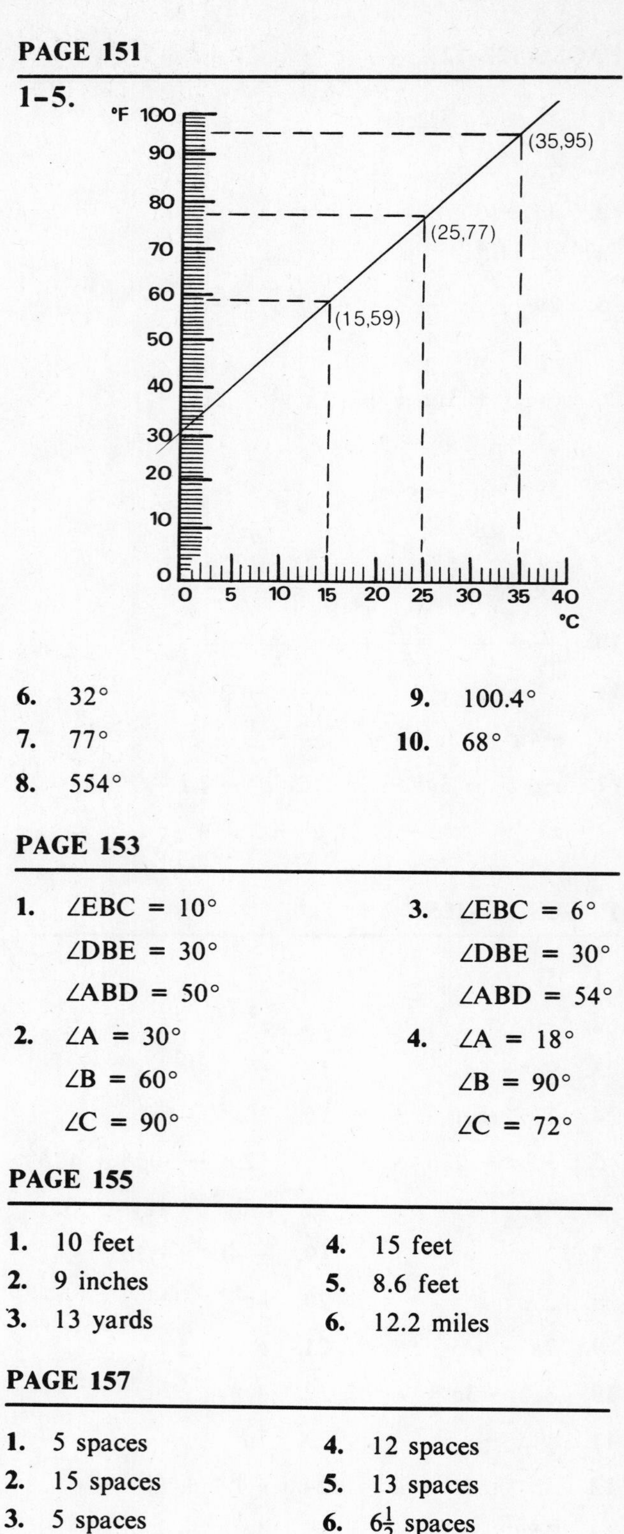

6. 32°
7. 77°
8. 554°
9. 100.4°
10. 68°

PAGE 153

1. $\angle EBC = 10°$
   $\angle DBE = 30°$
   $\angle ABD = 50°$
2. $\angle A = 30°$
   $\angle B = 60°$
   $\angle C = 90°$
3. $\angle EBC = 6°$
   $\angle DBE = 30°$
   $\angle ABD = 54°$
4. $\angle A = 18°$
   $\angle B = 90°$
   $\angle C = 72°$

PAGE 155

1. 10 feet
2. 9 inches
3. 13 yards
4. 15 feet
5. 8.6 feet
6. 12.2 miles

PAGE 157

1. 5 spaces
2. 15 spaces
3. 5 spaces
4. 12 spaces
5. 13 spaces
6. $6\frac{1}{2}$ spaces